RESEARCH IN MARITIME HISTORY
NO. 26

IN TROUBLED TIMES: THE PORT OF LIVERPOOL, 1905-1938

Adrian Jarvis

International Maritime Economic History Association

St. John's, Newfoundland
2003

ISSN 1188-3928
ISBN 0-9730073-6-2

Research in Maritime History is available free of charge to members of the International Maritime Economic History Association. The price to others is US$15 per copy, plus $3.50 postage and handling.

Back issues of *Research in Maritime History* are available:

No. 1 (1991) David M. Williams and Andrew P. White (comps.), *A Select Bibliography of British and Irish University Theses about Maritime History, 1792-1990*

No. 2 (1992) Lewis R. Fischer (ed.), *From Wheel House to Counting House: Essays in Maritime Business History in Honour of Professor Peter Neville Davies*

No. 3 (1992) Lewis R. Fischer and Walter Minchinton (eds.), *People of the Northern Seas*

No. 4 (1993) Simon Ville (ed.), *Shipbuilding in the United Kingdom in the Nineteenth Century: A Regional Approach*

No. 5 (1993) Peter N. Davies (ed.), *The Diary of John Holt*

No. 6 (1994) Simon P. Ville and David M. Williams (eds.), *Management, Finance and Industrial Relations in Maritime Industries: Essays in International Maritime and Business History*

No. 7 (1994) Lewis R. Fischer (ed.), *The Market for Seamen in the Age of Sail*

No. 8 (1995) Gordon Read and Michael Stammers (comps.), *Guide to the Records of Merseyside Maritime Museum, Volume 1*

No. 9 (1995) Frank Broeze (ed.), *Maritime History at the Crossroads: A Critical Review of Recent Historiography*

No. 10 (1996) Nancy Redmayne Ross (ed.), *The Diary of a Maritimer, 1816-1901: The Life and Times of Joseph Salter*

No. 11 (1997) Faye Margaret Kert, *Prize and Prejudice: Privateering and Naval Prize in Atlantic Canada in the War of 1812*

No. 12 (1997) Malcolm Tull, *A Community Enterprise: The History of the Port of Fremantle, 1897 to 1997*

No. 13 (1997) Paul C. van Royen, Jaap R. Bruijn and Jan Lucassen, *"Those Emblems of Hell"? European Sailors and the Maritime Labour Market, 1570-1870*

No. 14 (1998) David J. Starkey and Gelina Harlaftis (eds.), *Global Markets: The Internationalization of The Sea Transport Industries Since 1850*

No. 15 (1998) Olaf Uwe Janzen (ed.), *Merchant Organization and Maritime Trade in the North Atlantic, 1660-1815*

No. 16 (1999) Lewis R. Fischer and Adrian Jarvis (eds.), *Harbours and Havens: Essays in Port History in Honour of Gordon Jackson*

No. 17 (1999) Dawn Littler, *Guide to the Records of Merseyside Maritime Museum, Volume 2*

No. 18 (2000) Lars U. Scholl (comp.), *Merchants and Mariners: Selected Maritime Writings of David M. Williams*

No. 19 (2000) Peter N. Davies, *The Trade Makers: Elder Dempster in West Africa, 1852-1972, 1973-1989*

No. 20 (2001) Anthony B. Dickinson and Chesley W. Sanger, *Norwegian Whaling in Newfoundland: The Aquaforte Station and the Ellefsen Family, 1902-1908*

No. 21 (2001) Poul Holm, Tim D. Smith and David J. Starkey (eds.), *The Exploited Seas: New Directions for Marine Environmental History*

No. 22 (2002) Gordon Boyce and Richard Gorski (eds.), *Resources and Infrastructures in the Maritime Economy, 1500-2000*

No. 23 (2002) Frank Broeze, *The Globalisation of the Oceans: Containerisation from the 1950s to the Present*

No. 24 (2003) Robin Craig, *British Tramp Shipping, 1750-1914*

No. 25 (2003) James Reveley, *Registering Interest: Waterfront Labour Relations in New Zealand, 1953 to 2000*

Research in Maritime History would like to thank Memorial University of Newfoundland for its generous financial assistance in support of this volume.

Table of Contents

List of Figures

Acknowledgments

The research for this book has been carried out over a period of about three years, during which time I have presumed on the goodwill of a lot of people, including the staff of Liverpool University's Harold Cohen and Sydney Jones Libraries and Mike Chrimes, Head Librarian, and Carol Morgan, Archivist, at the Institution of Civil Engineers. My greatest debt in this direction is, as ever, to my colleague Dawn Littler, Curator of Archives at Merseyside Maritime Museum and her staff, who patiently put up with me wandering in and out of their offices and stacks. My privileged position in that respect has allowed me to employ the invaluable, if under-rated, research technique of serendipity: the best bits of evidence are always the ones you find by accident. It has also saved me a vast amount of time.

In one chapter where I did not trust my own judgement and another where I was doubtful about my contextual knowledge I sought the advice of Professor Sarah Palmer and Professor John Armstrong, respectively, who kindly read and commented on the two chapters in question. I claim sole intellectual property in any errors which remain.

Among the many benefits of my position as virtually a full-time research historian is the ability to get around and meet people at conferences and elsewhere, and this book would have been much less adequate than it is (a subject on which opinions will doubtless vary) without the innumerable hints and snippets that one picks up in passing, to say nothing of more substantial insights or pieces of advice. I cannot possibly name all the people involved and it would be invidious to do so even if I could, but I am sure that the rest will forgive me if I mention Dr. Gordon Jackson, the man who seized the history of ports back from the geographers for the historians and gave it a philosophical framework. Nobody who works in the field can fail to be influenced by him or to benefit by picking his brains. Among my colleagues in Liverpool, Mike Stammers, Keeper Emeritus, Merseyside Maritime Museum, has provided extensive information and references (especially on flats and barges), and Professor Peter Davies has given me the benefit of his long experience of the world of the shipowners. Professor Lewis Fischer's work on port efficiency has suggested a methodological framework for investigating the thorny problem of port investment which I hope adds something to my attempts at evaluating the success or otherwise of the Port of Liverpool in meeting the challenges of an extremely troubled period.

I have a long-term voluntary research assistant named Richard Foyan who comes and crunches numbers for me a couple of hours a week. He has patiently ground his way through many calculations, especially for present purposes those used in the appendix, and I am greatly indebted to him, especially for his cheerful acceptance of the fact that the issues we try to determine usually turn out to be too complex for the figures to be conclusive.

Finally, I like to think that others than me will continue to benefit from the policy pursued by the Trustees of National Museums Liverpool ever since their inception, of supporting research into the Museums' collections and the publication of the results. History is a cumulative discipline in which information-based institutions cannot stand still: it is not just their display and communication methods which should be as good as they can be made through constant improvement but the information content itself as well.

Conversion Factors for Imperial Measurements

Everything described in this book was designed and built according to the Imperial System and paid for in pre-decimal money. In the interests of easier reading, conversions into metric, SI and decimal currency have not been interpolated in the text. For those unfamiliar with the old systems, the following conversion factors are offered.

1. Money
The old system divided the pound sterling into 20 shillings (s) and the shilling into 12 pence (d). In some cases, such as rates of wages, it was conventional *not* to reduce shillings to pounds, so that a well-paid craftsman might be described as receiving 55s, i.e., £2.75p per week.

1 shilling = 5 new pence.
1 old penny = 0.42 new pence.

2. Distance
1 mile = 1760 yards = 1609 metres.
1 yard = 3 feet = 0.914 metres.
1 nautical mile = 6080 feet = 1855 metres.

3. Heights and Depths.
Are normally measured in feet and inches above or below the level of the sill of Liverpool's first dock, normally known as Old Dock Sill, occasionally Old Dock datum. This lies 4 feet 6 ½ inches below present-day Ordnance Datum.
1 foot = 12 inches = 0.305 metres.
1 inch = 25.4 mm.

4. Area
1 acre = 4840 square yards = 0.405 hectares.
1 square yard = 9 square feet = 0.836 square metres.
1 square foot = 0.304 square metres.

5. Volume
Solids, such as spoil or masonry, are measured in cubes of linear measurements, while liquids (and quasi-liquids like bulk grain) are measured in bushels, gallons etc.
1 cubic yard = 27 cubic feet = 0.764 cubic metres.
1 bushel = 8 gallons = 36.4 litres.

1 gallon = 8 pints.

6. Weight
1 ton = 20 hundredweight (cwt) = 1.016 tonnes.
1 cwt = 112 pounds (lbs) = 50.8 kilograms.

7. Pressure
1 pound per square in (lb/in^2) = 6.895 kilopascals = 0.0703 kg/cm^2.

8. Power
1 horsepower = 33,000 ft/lb per minute = 0.746 kilowatts.

Introduction: The Port of Liverpool in 1905

In 1905 Liverpool was still by a wide margin the largest port in the country administered by a single port authority, the Mersey Docks and Harbour Board (MD&HB). It was the second port in the Empire in terms of the value of goods handled and Britain's largest export port. This situation resulted from nearly two centuries of investment in the infrastructure of the port, beginning with the Liverpool Corporation's decision in 1709 to seek an Act of Parliament to enable the construction of what was to be the first commercial wet dock in the world.

This was necessary because, as numerous writers have remarked, any advantage Liverpool might derive from its favourable geographical position was nullified by the hostile and often dangerous nature of its river.[1] Samuel Smiles and his imitators often made such statements as "the task of the engineer is to transform," by which they meant that what Mother Nature gave with one hand, she might well take away with the other. Liverpool, like Bristol, could only respond to economic stimuli, grow and prosper through the ingenuity of the engineers - and the raising of money to pay for their work. Each time the engineers found a solution, the port's customers altered the question, a situation which remains fundamentally unchanged between 1709 and the present.

The result was a continuum of dock engineering work which might be considered in three categories: repairs and minor improvements; major remodelling of old works; and construction of new works. The first were normally funded from revenue, principally dues on goods and ships, and the last invariably from borrowed capital, with the intermediate category funded either way, according to cost and the state of the port authority's cash flow.[2] Since this led to a continuing growth of debt, the only way to continue modernising while meeting

---

[1]As I write these words, the ebb tide is running at about five knots: through my window I can see Dukes Buoy heeled over about twenty degrees and moving up and down by about six feet in the grasp of one of the Mersey's characteristic "mini-swells." A small bunkering tanker is labouring very slowly upstream and sheets of spray are breaking over the end of the floating stage. Yet this is not an especially big tide, and there is only a stiff breeze blowing.

[2]Cash flow might well be governed by factors completely beyond the control of the port, such as prevailing rates of interest. The corporate debt was originally entirely in "Dock Bonds" - effectively mortgages on future dock dues - but by 1905 there were also lesser sums raised by the issue of annuities and debenture stocks. For further details, see chapters one and five.

interest payments was through the continuing expansion of trade, which implied an expansion of the physical extent of the port.[3] As a result, by 1905 the water area of the Liverpool docks had risen from the 3.5 acres of the first dock to 336 acres, while across the river at Birkenhead a further 160 acres had been added. This arca was in thc process of increasing still further, with major works in hand on both sides of the river, and plans were being discussed for a further wave of expansion once the works under construction were complete.

In 1893 the Liverpool Overhead Railway was opened, running the full length of the Dock Estate from Herculaneum Dock in the south to Hornby at the north, a distance (by rail) of some 5.75 miles, with the docks themselves extending a little further at both ends (for the location of the various docks, see figures 1a, 1b and 1c on pages 7-9). Although the railway had been built to carry people having business on the docks, it rapidly acquired a quite different additional clientele: people who wanted to see what happened in that hidden world inside the high boundary walls of the Dock Estate. Let us gain an idea of the extent and condition of the port by taking an imaginary trip along the railway.

Herculaneum Station, where we begin, provides a panoramic view of Herculaneum Dock, originally opened in 1866 but modernised in 1881 and extended by the addition of a branch dock, which is now normally occupied by tankers discharging bulk petroleum products to the Parkhill depot, which is visible beyond and above the dock.[4] Along the landward and south sides of the dock is much higher ground, where the "casemates," little man-made caves with massive iron doors, are dug back into the rock. These were Liverpool's first purpose-built petroleum storage facilities, back in the days when oil came in barrels. Also on the east quay are two coaling cranes, mainly for bunkering rather than export coal. Along the south side are four graving docks, one of them newly opened, varying in length from 754 to 930 feet, and in entrance width from sixty to eighty feet. They are not, however, designed for particularly large vessels but rather to each take two or three more modest ones.

Over on the river wall is a fairly prominent workshop complex belonging not to the Dock Engineer's department but to the Marine Surveyor's: its mission is the repair and maintenance of buoys, and adjacent to it is the pumping station for the graving docks, ingeniously arranged so that it can also pump river water

---

[3]Note the use of the word "implied" rather than "mandated." It was theoretically possible to increase trade without increasing extent through improved efficiency: a reduction in ship turnaround time, for example, saves berth space. In practice, however, the pattern of growth was also a continuum of obsolescence, which rather militated against (though did not completely preclude) significant improvements in efficiency.

[4]Bulk crude has never been imported on the Liverpool side of the river; see chapter seven.

into the wet dock to provide extra depth when needed. Directly across the road from the railway station are the two-story transit sheds of Harrington Dock, with travelling hydraulic roof cranes. Like Toxteth, the next dock to the north, Harrington was part of the last really large-scale new construction programme, enabled under an Act of 1873, and they were completed in 1883 and 1888, respectively. Designed for the medium-to-large cargo liners used by firms like T. and J. Harrison and Elder Dempster, they have proved quite successful despite the high initial cost of the sheds. To the right of the Overhead Railway is the first of many mainline railway facilities on the inland side; this one belongs to the Cheshire Lines Committee.

Travelling northwards past the tiny Union Dock, which is little more than a passage, the next facility is Brunswick, opened in 1832 as a specialist timber dock. It now serves the general cargo trade - or it will do once the huge works currently underway are completed. Other users have been removed from the site, and a large new river entrance is virtually complete. The old entrance and its half-tide basin have been left in, for north of them stands the Dock Yard - the principal depot of the Engineer-in-Chief - with its complex of office and workshop buildings tightly packed onto a site roughly 400 yards square; it is still the largest establishment of its kind in the world. The old half-tide dock is used by the Engineer's floating plant - dredgers, hopper barges, floating barge cranes, salvage "lumps" and the like, as well as by vessels in need of repair owned by other departments of the Board.[5] One vessel not there is the Board's sand-pump hopper dredger *Coronation* - at 3500 tons the largest dredger in the world - which is working on the approach channel to the new Brunswick entrance.[6]

Continuing north to Coburg Dock we find minor deepening works being completed. Queens Dock, a little further on, has been modernised beyond all recognition and splendid two-story sheds with electric lighting are just nearing completion, as is the large new graving dock which will replace several older and smaller ones. On again to Kings Dock where the scene is very similar. A spacious but decidedly shallow eighteenth-century dock has been transformed from being almost square to having a "spine" incorporating Wapping Dock and two branches. All this work has taken or is taking place under the "New Works" Act of 1898, and it is already encouraging other investors: at Coburg, for example,

---

[5]The Board's fleet was quite extensive by this time; other vessels included the warehouse department's steam barges *Tamar* and *Tay,* the harbourmaster's tugs *Hodgson* and *Neptune*; and the marine surveyor's watch vessels and tenders, including the passenger tender *Galatea*. Still in service was the veteran schooner *Oak,* designed and built in the Dock Yard in 1836 to fetch granite from the Board's quarry at Kirkmabreck. For further details, see chapters six and seven.

[6]This is the carrying capacity, the customary figure to quote for hopper dredgers.

a group of large grain silos is under construction for the Liverpool Grain Storage and Transit Company. What we have seen so far fully lives up to the idea of a continuum of modernisation.

But our imaginary train has carried us only about 2000 yards, and as it leaves Wapping Station the landscape changes. We see Duke's Dock, built for the Duke of Bridgewater in 1773 and served by a warehouse built in 1783. Everything about it is boat- rather than ship-scaled, and because it was until recently an outpost of Manchester, not part of the Port of Liverpool, it has its own jumble of workshops, stores and offices crammed into every corner of the site. There is Salthouse Dock, opened in 1753 and modernised in three phases, the last as recently as 1855. It has ill-lit, narrow, single-story transit sheds and very narrow quay margins, and the amount spent on its maintenance this year is zero. Beyond, on the river front stand the mighty Albert Dock warehouses, still turning a profit as warehouses but with virtually no ships around small enough to use the dock's entrances. A forty-five-foot passage with about twenty-one feet on the sill at mean high water was quite generous in 1846 but is almost useless now: the goods mostly come and go by road, rail and barge. North again to Canning Dock, Canning Half-tide and Canning Graving Docks, basically still as Jesse Hartley left them in 1842. A few steam trawlers, coasters and vessels waiting for the graving docks are all they hold now. On the river front are two small basins for barge transfer traffic, Manchester Dock and Chester Basin. Manchester Dock has one quite smart-looking little warehouse, but only because the old one burned down in 1890.

Just across James Street from this scene of decay is a construction site. The building rising on it is the new Dock Offices being built for the MD&HB at a cost of £300,000; it is already clear that it will be a building of considerable size and splendour. But it would also be noticeable to those privileged to see the drawings that all the important people have rooms with a view over the river, not over the decay we have just witnessed. The site on which it stands is part of the former Georges Dock, built in 1771 and last modernised in 1822-1825, but filled in by 1900.

Our tour has so far shown us three excellent docks, three fairly good ones and a lot of obsolete wasted space. This is most curious. The former Dock Trustees were among the pioneers of the "corporate image" in Liverpool, with their magnificent office at Albert Dock (still standing, and now housing a Granada TV studio), their scarcely less grand Harbourmaster's Office on the corner of Mann Island, and Jesse Hartley's sorties into neo-medieval architecture with his pump houses and accumulator towers. The MD&HB had maintained the policy with Lyster's pump houses, the pilotage building and, above all, the new offices now being built. Yet on its own doorstep it allowed this public display of obsolescence to continue.[7] It would be a difficult area to modernise, but its value could be phenomenal.

To the north of Pierhead the situation improves, but not by much. Princes Dock dates from 1821, but has been tolerably well adapted to provide decent berths for the Irish trade, and in 1905 work is in hand to widen the quay margins on the west side by constructing ferro-concrete stagings out into the water. Along half its length a new transit shed of generous width is being built, again partly over the water to fit it into the very cramped site. The further end of the west quay and all the other side of the dock are still somewhat antiquated, but the dock in fact still attracts considerable trade. The half-tide entrance and dock are more effective than they look at first sight, and unlike any of the docks described above as obsolete, Princes has been equipped with an hydraulically-operated passage bridge. What could not be seen from the train was the arrangement for pumping out the little Princes Graving Dock, consisting of an 1838 railway locomotive mounted on blocks with its fly cranks and coupling rods removed and a bevel gear pressed onto its crank axle to provide a power take-off point. Heath Robinson would not have dared suggest such an arrangement.[8]

Beyond the river wall, and extending from James Street station to Princes Dock station, is one of Liverpool's most evocative assets, the Princes Stage, the longest floating structure in the world. Here dozens of ferryboats ply their trade on the estuary, interspersed with larger ferries serving such places as North Wales, the Isle of Man and Glasgow. But what creates the glamour is, of course, the ocean-going passenger vessels, be they the comparatively modest Elder Dempster liners bound for West Africa or the monsters of the North Atlantic. The stage has its own railway station, and Princes Parade, which lies between the station and the stage, has a glazed roof to spare the cabin-class passengers from unpleasant weather while walking about fifty yards to their ship.

Proceeding further north we find the shallow and obsolescent East and West Waterloo docks. East Waterloo has three blocks of bulk-grain warehouses which were world leaders in their day, but their day was 1868. West Waterloo, after having been home to some of the biggest sailing vessels using the port, is now busy with smaller craft. Victoria and Trafalgar have scarcely been modernised at all since Hartley finished them in 1836 and are suitable only for the coastal trade. Then we arrive at the extraordinary time-warp which is Clarence Dock. The strange thing about Clarence is not so much its survival virtually unchanged since 1830 as that it still carries considerable traffic, with coasters

---

[7]For performance figures of individual docks, see chapters one and *seriatim*.

[8]The locomotive was *Lion*, built for the Liverpool and Manchester Railway in 1838, donated by the MD&HB to the Liverpool Engineering Society for preservation in 1926 and now in the National Museums Liverpool (NML).

coming and going on every tide. It is, however, a disgrace to the modern port.[9] North again are Salisbury and Collingwood, opened in 1848 and barely altered since. Salisbury has a double half-tide river entrance with a handsome clock tower on the island, while Collingwood was the last dock purpose-built for the coastal trade. These, too, are thronged with coasters but produce little revenue.

To the right is Stanley Dock, the only one to be excavated entirely from dry land. Some might think it had been built in an unwise place to start with, inland of the dock railway, the dock road and (since 1893) the Liverpool Overhead Railway. Here is a major new investment in the shape of the biggest tobacco warehouse in the world, opened in 1901[10] – served by a dock with a fifty-foot entrance obstructed by a swing bridge which can only open at night when the trains have stopped. Its use by shipping is very small – an aggregate tonnage of 14,969 in 1905 – and the warehouse strangely cuts off its own railway access by being built to the last inch that would fit on the site; it has yet to be re-connected from the other side. Its very large traffic (for it is enormously successful) largely comes by transfer barge and leaves by road.[11] It should not be forgotten, however, either that Liverpool is at this time a leading centre in the development of steam road lorries or that A.G. Lyster, the Board's Engineer-in-Chief, is a man both unenamoured of railway companies and unfrightened by innovation.[12] Since Liverpool was then a major centre of tobacco processing (and still has the largest pipe tobacco factory in Britain), a good deal of the outward traffic was short-haul.

---

[9]For further details, see A. Jarvis, *Liverpool Central Docks 1799-1905. An Illustrated History* (Stroud, 1991), 208-211.

[10]The author is well aware that the date on the building itself is 1900, but the *Engineer's Report* of July 1901 describes it only as "practically complete." The date plaque was obviously specified in the same spirit of optimism as the medal struck to commemorate the Third Reich's victory in North Africa (an example survives in the King's [Liverpool] Regiment Collection, NML).

[11]Barge movements are extremely difficult to trace (see chapter seven), and the day-to-day records of the building have not survived.

[12]The Liverpool Steam Vehicle Trials of 1899-1900 attracted nationwide attention. Liverpool University's first Professor of Engineering, H.S. Hele-Shaw, was a prominent figure in early steam and motor transport by road, and papers on the subject appeared in the *Transactions of the Liverpool Engineering Society*. For Lyster on railway companies, see his "Presidential Address," *Minutes of the Proceedings of the Institution of Civil Engineers,* CXXV (1913-1914), 4-41.

Figure 1a
Liverpool – The North Docks

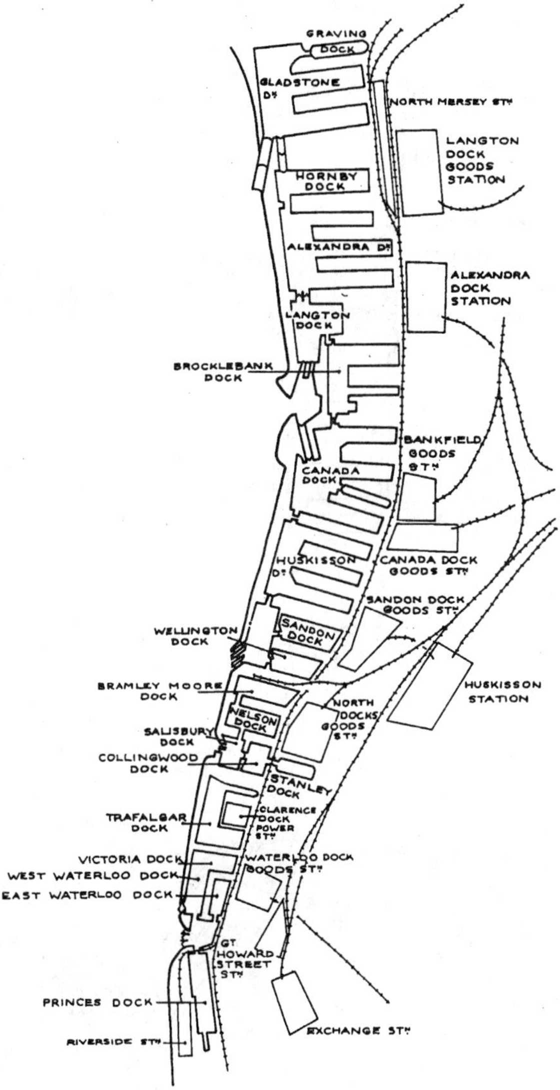

Source: John Hughes, *Port in a Storm. The Air Attacks on Liverpool and Its Shipping in the Second World War* (Liverpool, 1993), 179-181.

Figure 1b
Liverpool – The South Docks

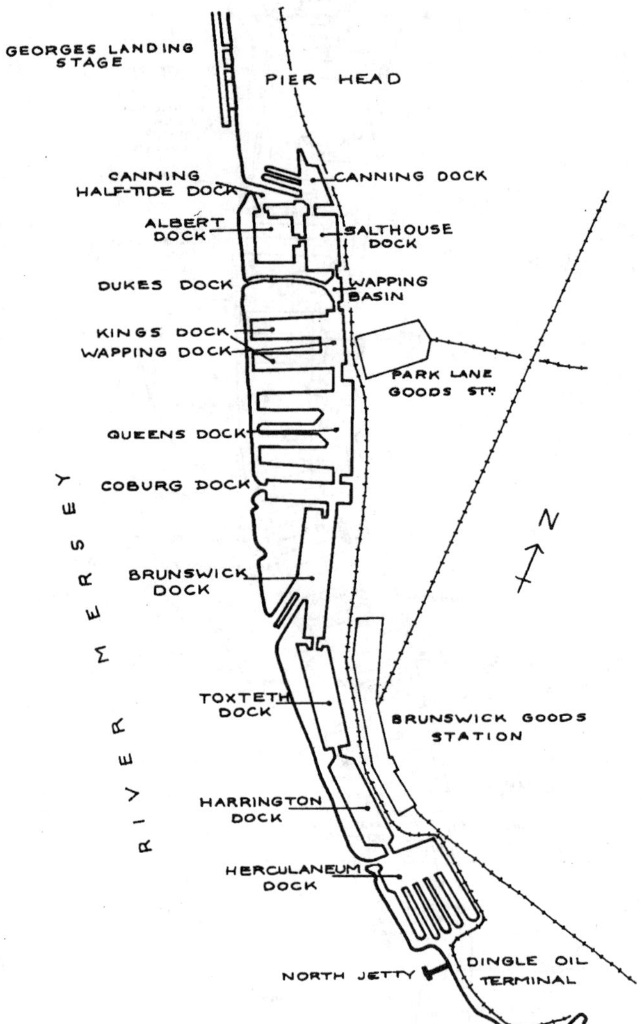

Source: See figure 1a.

Figure 1c
Birkenhead Docks

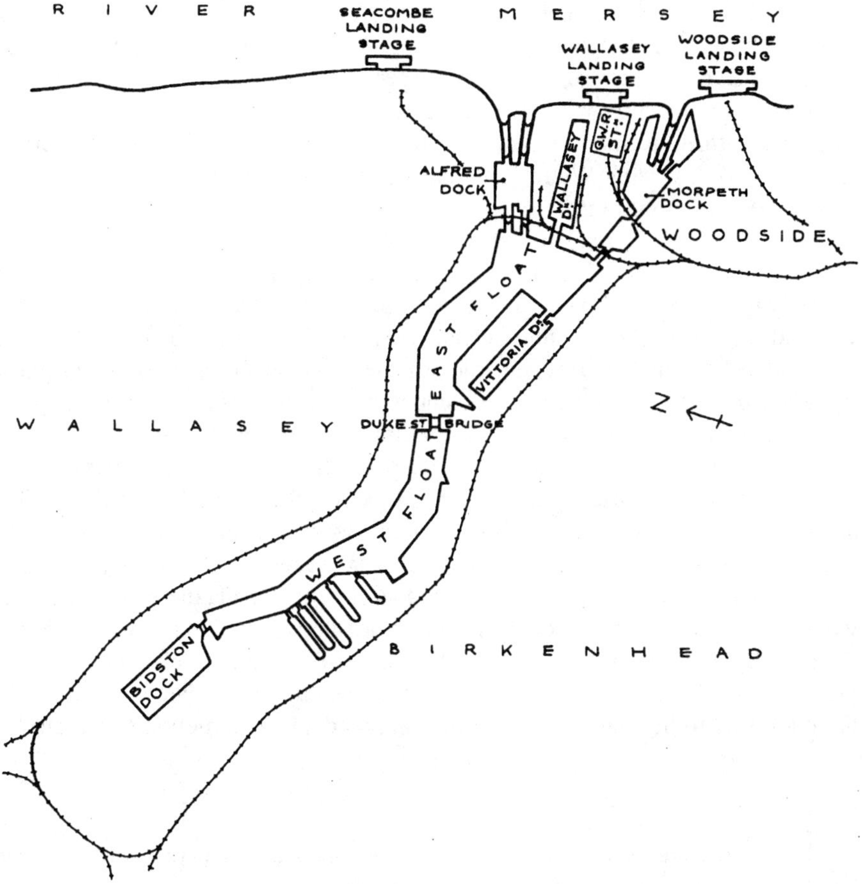

Source: See figure 1a.

Passing Nelson Dock, we come to Bramley-Moore, where the Overhead dips briefly to ground level to avoid the High-Level Coal Railway, which comes across from the Lancashire and Yorkshire Railway on a viaduct which runs along the heads of both Bramley-Moore and Wellington docks, with a branch down the north quay of Bramley-Moore. The idea was that comparatively modest cranes could empty railway wagons into holds or bunkers, but of course as ships got deeper they also became taller when unladen, and now most of the coal trade here is in filling floating elevators and their attendant barges for coaling large vessels in the river or in deeper docks. The rest of Bramley-Moore and Wellington, as well as Nelson, is chiefly occupied by vessels trading to mainland Europe and the Mediterranean, though some of the larger coasters appear as well. Beyond Wellington on the river side we find the first evidence of handling the really big ships in the form of the new Sandon Entrance, which has obliterated the former separate basins serving Wellington and Sandon docks under a large new basin with a deep-water entrance.

As we reach first Huskisson and then Canada Docks we see the ships the tourists come to admire – in Huskisson there are appropriated berths for Ismay Imrie, which presently owns the four largest ships in the world, Cunard, the Leyland Line and others, while in Canada are further Ismay Imrie berths along with Allan Line and more companies of the Leyland group.[13] Handling these vessels provides immense revenue compared with the coastwise and sea-going trades, and the scale and quality of the facilities provided makes that obvious. The Langton graving docks are no longer deep enough for the largest vessels, but Langton Branch is in use by the Ellerman Line and the ubiquitous Leyland Line, always among the largest contributors to the Board's coffers. Between the two graving docks stands the amazing mock-Bavarian castle which is the pump house.

The further we go, the bigger things get. At Canada there are deepening works in hand and a branch dock is under construction, while the obsolete South Carriers Dock has been obliterated by the almost-complete Brocklebank Graving Dock. In Brocklebank Dock itself, further deepening works can be seen. In all, there are said to be some 4000 men working on dock construction at present.[14]

---

[13]An appropriated berth is one for which a shipping line pays a rent in order to secure preferential use.

[14]There is some doubt about most of the figures quoted in the records, and the mix of the Board's "extra" and "established" men and the contractors' employees makes the establishment of a completely reliable figure unlikely. This estimate from Great Britain, Parliament, *Parliamentary Papers* (*BPP*), 1902 (1151), XLIII, 222, evidence of Miles Kirk Burton, General Manager and Secretary of the MD&HB, sounds reliable, but other parts of his testimony were demonstrably false. This issue recurs in the accounts of construction of specific docks below.

The last two docks, Alexandra and Hornby, are now in sight, opened in 1880 and 1884, respectively. Such is the pace of change that their revenue is already beginning to slide, as primacy is taken from them by the deeper entrance and berths of the Canada-Huskisson improvements. Still, they remain large, modern, impressive and busy, home to such firms as Canadian Pacific and Pacific Steam Navigation Co. But to the north of Hornby we see that the Engineer's Department can never rest, for the removal of a spoil bank left from previous works is nearing completion: this is part of the site of a new dock "in contemplation," as they put it, for 1000-foot vessels.

How the Port Authority Worked

Everything we have seen was ultimately the product of decisions made by the port authority, and it is to a brief consideration of how this body functioned that we must move next. When the first Act was obtained in 1709, the councillors and aldermen of the Common Council were appointed Trustees of the Liverpool Dock, a situation that remained unchanged throughout the eighteenth century apart from the gradual emergence of a Dock Committee to which the powers and duties of the Trustees were effectively delegated, but subject always to the confirmation of the "The Trustees of the Liverpool Docks in Council Assembled." By 1793 the Dock Committee gained its own Minute Book. Until 1811 there was really no check at all on the power of the Trustees to do as they wished apart from that inherent in the need to gain Parliamentary powers to raise capital or make compulsory purchases of land. Under the 1811 Dock Act a new system of auditing was introduced which for the first time provided customers with some slight "come-back" if the Trustees abused their powers. When the Act of 1825 allowed for the election of "dock ratepayers" (those who paid dues on ships or goods using the port) to the Dock Committee, the number of non-councillors increased gradually until the membership stood equal at twelve council members and twelve elected members - but the Chairman, who held a casting vote, had to be a councillor. The situation remained this way until the upheaval of 1857.

Meanwhile, the growth of the port and its trade, and the complexity of everything to do with them, had begun to overtax the Dock Committee. This is scarcely surprising if we recall that between 1793 and 1847 the number of vessels paying dues rose from 4129 to 20,889, and the amount they paid rose from £12,480 to £244,435.[15] The mid-1840s were marked by astonishingly large and rapid construction, and to keep closer control over the principal officers the Dock Committee in 1846-1847 established a system of sub-committees, each responsi-

---

[15]The main reason for the much greater proportionate increase in revenue was, of course, the growth in the average size of ships using the port. But there was also a tendency towards higher-value cargoes, which paid larger dues.

ble for one distinct area of their activities; these initially were Finance, Works, Department of Harbour and Dockmasters, Department of the Marine Surveyor and Warehouse Management. In 1851 ongoing conflict over priorities in the discharge of cargoes led to the creation of a sub-committee for Master Porters.

Despite being a more or less self-perpetuating oligarchy for most of the time, the old system served Liverpool well. Its members were able men who were supported by some outstanding principal officers. They were, in fact, too good: the amount of revenue they generated and the ruthlessness with which they eliminated competition gained them a succession of enemies who, when they combined, proved powerful enough to get a Bill through Parliament which literally dispossessed Liverpool of its docks and established a non-profit statutory trust in which were vested not only the physical assets but also effectively all the rights and duties of the former body of Trustees. Membership of the Board was (eventually) set at twenty-eight, of whom twenty-four were to be elected by port users and four nominated by the Crown (represented by the Acting Conservator of the Mersey). The Committee system was altered in detail only.

The question of whether the Board, which held its first meeting on 5 January 1858, was a better or a worse way of running a port is one on which there is already a considerable literature and which need not be reconsidered here.[16] For the purposes of this brief introduction it is sufficient to remark that the constitution of the Board was something of a mixed blessing, that it was widely imitated elsewhere and that it proved remarkably durable, surviving with only a little "fine tuning" and evolution of its unwritten conventions, until its total financial collapse in 1970-1971 and its re-construction as the Mersey Docks and Harbour Company.

One of the unwritten intentions behind the constitution of both the old body and the new was that the operation of the port should be more closely connected with the rest of the local business and manufacturing community. Success in this was at best partial, degenerating rather than improving over the course of the nineteenth century. Part of the problem was clearly the perceived hierarchy of ways of earning money: shipowners and merchants might have their differences (and often did), but on one thing they agreed: they were definitely a cut above those in trade and manufacturing. Little work has yet been done on these people, their social contacts or their business networks for the period with

---

[16]For a summary of this literature, see A. Jarvis, "Managing Change: The Organisation of Port Authorities at the Turn of the Century," *The Northern Mariner/Le Marin du nord*, VI, No. 2 (April 1996), 31-42.

which this book is concerned.[17] Many of them qualify for the description "elite" by virtue of having left six or even seven-digit sums in their wills or by the style of life we can tell they led by looking at what survives of their housing. If we compare a local aristocratic home like Croxteth Hall (seat of the Earls of Sefton) and a successful shipowner's home like Rankin Hall on the edge of Sefton Park, there is little difference in the standard of architecture, construction or decor. Rankin's entertaining rooms are only slightly smaller, and the functions of the principal rooms - anteroom, dining room, ballroom, billiard room, etc. - are separated in similar ways. The main difference is in the number of rooms (Croxteth, like many stately homes, has a couple of dozen guest rooms and suites, for example) and the amount of land. The essential point is that Rankin's house was designed to announce that it was, as auctioneers put it, "the property of a gentleman."[18]

If we look at the main and subsidiary occupations of members of the Dock Board, we find that a high proportion were in commerce or shipowning, with a significant number having secondary interests in banking or insurance.[19] They quite clearly looked down on the majority of members of the Borough, later City, Council and repeatedly refused to allow any acknowledged Council representation on the Board. Conversely, few members of the Board sought election to Council.[20] A.G. Lyster summarised this quite nicely when explaining why local authorities should not run ports: among their members there might be people like hoteliers who had a positive interest in slowing down ship turnaround. It went without saying that anyone "in trade" would succumb to the temptations of conflict of interest, which the gentlemen of the Board would never do. Later, a more expansive passage in J.S. Allan's essay "The Trade and Commerce of Liverpool" explained that "the Municipal Government is, to a large extent, separate from the the large shipping and overseas commercial interests and is mainly in the hands of the energetic men who control the large retail trades and

---

[17]There is, of course, good work for earlier periods, notably G.J. Milne, *Trade and Traders in Mid-Victorian Liverpool: Mercantile Business and the Making of a World Port* (Liverpool, 2000).

[18]There are those who might suggest that its Gothic style proved exactly the opposite, but the Duke of Westminster could build Eaton Hall without being dubbed *nouveau riche*.

[19]For an analysis of their interests, see chapter four.

[20]The forty-two page appendix of biographical notes in P.J. Waller, *Democracy and Sectarianism: A Political and Social History of Liverpool 1868-1939* (Liverpool, 1981), contains only a handful of MD&HB members.

other necessary works and occupations to be found in all great cities."[21] The choice of the words "energetic" and "necessary" is masterly in its disparagement.

Perhaps more surprisingly, there is little cross-membership of the Board and the Chamber of Commerce. This might be thought to result from the Chamber's sometimes acid complaints about the level of service provided by the Board to its members, but exactly the same explanation could be offered in the case of the Liverpool Steamship Owners Association, where it is not needed because there *was* extensive cross membership. It was, indeed, a key objective of the Association. The Board was often criticised for being aloof, and the mighty perimeter walls of the Dock Estate, together with the denial of public access to most of the length of the river front, were said to symbolise their attitude. How fair such accusations were is not necessary to discuss here: all we need is to identify some sort of separation, and there seems to be plentiful *prima facie* evidence of that.

Life and Work in Edwardian Liverpool

If the port authority could stand largely aloof from city politics, it could not detach itself from the city, for it was the interaction between the activities within the dock estate and the necessary services outside it - anything from marine insurance to the supply of marine glue - which turned the docks into a port. Conversely, almost all the trade and industry within the city was, in one way or another, driven by the ships and goods handled in the port. This in turn meant that most of the people in the city earned their living from something which depended directly or indirectly on the port.

Historians have taken sharply different views about what it was like to live in Edwardian Liverpool. To Ramsay Muir, followed by George Chandler, it was a city of wealth and splendour with a disreputable heritage of unhealthy living conditions but in which immense progress had been made and was continuing to be made.[22] To Eric Taplin, Frank Neal, Peter Waller and John Belchem the picture is much darker, exhibiting little improvement in half a century.[23] As we shall see, both views are correct, for Liverpool was home to

---

[21]A. Holt (ed.), *Merseyside. A Handbook to Liverpool and District* (Liverpool, 1923).

[22]J.R.B. Muir, *A History of Liverpool* (London, 1907); and G. Chandler, *Liverpool* (London, 1957).

[23]E.L. Taplin, *Near to Revolution: The Liverpool General Transport Strike of 1911* (Liverpool, 1994); F. Neal, *Sectarian Violence: The Liverpool Experience. An Aspect of Anglo-Irish History* (Manchester, 1989); P.J. Waller, *Democracy and Sectarianism: A*

great wealth and great squalor in remarkably close proximity. But before addressing that question directly, it is necessary to digress into the belief which underpins the arguments of the pessimists.

It is widely believed that Liverpool had an almost entirely commercial and service-based economy and hence a negligible manufacturing sector. The latter was *comparatively* small, and it would be foolish to claim that Liverpool was the equal of Glasgow in shipbuilding or of Manchester in mechanical engineering – but one would certainly never guess from existing literature that in the 1891 Census Liverpool had more skilled metalworkers (aggregated for all trades) than did Manchester, or that by the end of the nineteenth century the Port of Liverpool (including Birkenhead) was the second-largest flour-milling centre in the world. This latter cross-fertilised with its status as the third biggest vegetable oil milling centre in the country and the second biggest sugar refiner to make it a promised land to biscuit (and later boxed cake) manufacturers. In an earlier period, the supremacy of coastal shipping in the transport of heavy low-value goods like china clay made it, like Bristol, important for the manufacture of pottery and porcelain. Between 1796 and 1840, Liverpool's largest pottery, Herculaneum, occupied a site of 30,900 square yards in addition to its sixty-two houses and cottages for the workers. The site had formerly been used for a substantial copper smelting business.[24] Later, part of it was occupied by an extensive engineering works which in turn was closed to make way for dock improvements.[25]

In 1822 James Muspratt built what was easily the country's largest Leblanc alkali works in the town, reinforcing a local tradition of chemical production which was never going to be of national importance because so much of it migrated to Widnes and Runcorn, but which survived well into the twentieth century. Henry Sutton Timmis was one of a small number of chemical manufacturers who served as Board members. Manufacturing was not crucial, but over

Political and Social History of Liverpool 1869-1939 (Liverpool, 1981); and J.C. Belchem (ed.), *Popular Politics, Riot and Labour: Essays in Liverpool History 1790-1940* (Liverpool, 1992).

[24]A. Smith, *The Illustrated Guide to Liverpool Herculaneum Pottery 1796-1840* (London, 1970), 74 and 14-15, respectively. See also W.H. Chaloner, "Charles Roe of Macclesfield (1715-81): An Eighteenth Century Industrialist," *Transactions of the Antiquarian Society of Lancashire and Cheshire*, LXIII (1952-1953), 52-86.

[25]NML, Maritime Archives and Library (MAL), MD&HB, Legal, H44.

a long period of time it was far from negligible.[26] It may be that the emphasis on mercantile and service activities is in part a result of "victors' history."

Anywhere there was a livestock trade the size of that in Liverpool and Birkenhead there was going to be a tanning industry. A port that imported leaf tobacco in large quantities would have factories for pipe tobacco, cigarette manufacture and snuff milling, and so did Liverpool. It is all so simple and obvious – yet largely forgotten. Cope's cigarette factory (one of five in the city) was in a side street by Lime Street Station and was the better part of a quarter of a mile long, but who remembers it? On the contrary, we are told that one of the reasons for poverty in Victorian and Edwardian Liverpool was the lack of factory work to enable wives and children to augment the family income of underpaid and/or casually employed men in the transport sector, particularly dock labourers. This is an over-simplification which has only arisen because so little research has been done except in the history of the dock workers themselves. The exact number of men working as dockers over any particular period of time is not known, and probably never can be, but the potential supply was about 27,000, which is not a particularly large percentage of a population numbering over half a million.[27] The number of female jobs (excluding domestic service) was small, but again not negligible. Over 800 women worked as cardboard box makers; over 2700 toiled in laundries; over 1600 found employment in jam and sweet factories. The extensive sack and bag-making trade was dominated by women, and females were only just in the minority in ropemaking.[28]

But these are only the visible ones who appear in census data or studies like that by Marquis. These sources do not, for example, record part-time charwomen, and the office space necessary to house 16,535 general clerks must

---

[26]William Moss, *The Liverpool Guide* (Liverpool, 1796; reprint, Liverpool, 1974), largely avoids the subject of industry but lets slip, for example, that there was a large whale oil works (p. 25) and that "close by the prison are a steam mill for rolling and slitting iron and a White Lead Work." [sic] For other early industries, see R. Dickinson, "James Nasmyth and the Liverpool Iron Trade," *Transactions of the Historic Society of Lancashire and Cheshire*, CVIII (1956), 83-104; and A.E. Musson, *Enterprise in Soap and Chemicals. Joseph Crosfield and Sons, Limited, 1815-1965* (Manchester, 1965).

[27]Nor, incidentally, were dock labourers *necessarily* particularly badly paid: their problem was the casual nature of the work. When trade was really good the regular men could earn thirty-five shillings or more per week, far higher than the average for agricultural workers or railway servants. But actually working that many hours was unusual, and in lean years like 1908 there would be many for whom there was no work at all, with even the fortunate ones earning perhaps ten shillings for two days' work.

[28]Unless otherwise stated, all figures in this section are drawn from F.T. Marquis (ed.), *Handbook of Employments in Liverpool* (Liverpool, 1916).

alone have occupied quite a number of women. The passenger liners at the stage mostly employed shore-based cleaners who boarded them in dock to "blitz" the mess made by coaling before the ship tied up at the floating stage to collect the cabin-class passengers - and to put the cleaners ashore.

Other women made a small income from "self-employment" in such things as taking in washing or scrounging scrap wood and chopping and bundling it for kindling, and they, too, are absent.[29] On the fringes of the city markets was a sub-culture of small part-time street-traders and dealers in secondhand clothing, and, not being stallholders, they are also unquantifiable. The situation for women may not, therefore, be *quite* as black as has sometimes been painted, but it was certainly grim, and it was made worse by the fact that there was an acknowledged shortage of worthwhile jobs available to school leavers, denying poor families a small additional pay packet. It is true that of the 6000 or so messenger boys in the city a handful would in time be apprenticed as clerks, but for nearly all it was a dead-end job from which they got sacked at the age of sixteen, and the largest sources of male employment - dock work, road haulage and the railways - had only a very limited demand for boys.[30] Girls were possibly better off in terms of the number of openings in domestic service and dressmaking but much worse off in terms of wages. A few factory industries, including jam manufacture, employed a few girls at what was then a reasonable wage.

To moderate the view of the optimists it is only necessary to look at a large-scale Ordnance Survey (OS) map to realise that Liverpool contained a great deal of housing in which nobody would live by choice. We can roughly mark off plots in places like Rokeby Street, just north of the city centre, where blocks of four terraced houses had been squeezed onto plots about thirty-three-feet wide and sixty-six-feet deep - and then eight "back houses" fitted in behind them.[31] A particularly gross example on St. Anne Street reveals eighty-eight dwellings on a plot of little over 100 square feet, giving the unbelievable density of over 300 dwellings to the acre. Virtually the whole area from once-fashionable Everton

---

[29]Firewood-selling had a long history: the Dock Committee Minutes for 26 March 1837 record the provision of furnaces to burn scrap wood to prevent women scavenging for it on the dock estate. There is a passing mention of "chip chopping" in P. O'Mara, *The Autobiography of a Liverpool Irish Slummy* (Liverpool, 1934; reprint, Liverpool, 1972), 42.

[30]Colonel Henry Concanon, Manager of Ismay Imrie, started his working life in May 1871 at the age of fifteen as an MD&HB boy messenger, but not many others rose to such heights. In 1917 he was elected to the Board, becoming the first former employee to be thus honoured.

[31]For these pitiable little buildings it is, of course, necessary to use the ten feet = 1 mile series.

Brow down to the docks, well over a square mile, was occupied either by industry or by housing which at best consisted of exceptionally small terraced houses or the shabby ruins of once-pleasant larger terraces, often of courts.[32] This was the largest concentration of bad-to-disgraceful housing in Liverpool and a principal source of the gatherings of "five thousand ragged shoeless waifs" treated each Christmas Eve to "an ample Christmas dinner" in St George's Hall - but it was by no means the only one.[33]

Most of the principal office and shopping streets of central Liverpool varied between impressive and magnificent: this was a city where there was a great deal of money, both corporate and individual. Not for nothing was Bold Street known as "the Bond Street of the North." On Victoria Street, however, it was possible to turn a corner past a lavish office building and find oneself in a court of houses in which one would certainly not be allowed to keep animals today.[34] Around much of the city, appalling poverty quite literally lurked around the corner.

This, it should be remembered, was well into the age of "gas and water socialism," and Liverpool Corporation had struggled pretty boldly with its problems. In 1869 it had completed the first municipal housing in the country, St. Martin's Cottages, which were built for rent by the poor. They may have been spartan, but they had the unprecedented provision of a water tap and a water closet in each flat. They had rubbish chutes, good ventilation and drainage, were structurally sound and were not damp. Furthermore, since 1864 the Corporation had held standing powers of compulsory purchase to enable it to demolish "structurally insanitary" properties. But Liverpool's population was still rising, albeit at a slackening rate, and coupled with the enormous backlog existing in 1864, this meant that the number of grossly unfit properties only fell from about 20,000 in 1864 to 10,000 by 1900. It appeared to be an almost insoluble problem upon which repeated waves of major Corporation building had little impact because there were simply not enough new dwellings going up.[35] It is worth

---

[32]There is some doubt about the standard of these: largely built "on spec" by the close-knit community of Welsh builders, all one can say with certainty is that some builders were better than others and that the worst ones were dreadful.

[33]In 1896 the magazine *Sunday at Home* published a series of articles on Liverpool, and the one on pages 307-312 is on "The Children of the City."

[34]Victoria Street also housed, among other users, a fair proportion of Liverpool's 100 barristers' chambers.

[35]For early municipal housing in Liverpool and elsewhere, see J.N. Tarn, *Five Per Cent Philanthropy. An Account of Housing in Urban Areas between 1840 and 1914* (Cambridge, 1973); and Tarn, *Working Class Housing in 19th- Century Britain* (London,

recalling that the nightsoil boats on the Leeds and Liverpool Canal worked into the 1940s.

A partial solution was provided by the arrival of the municipal electric tramways, started in 1898. John Brodie, the City Engineer from 1898 to 1926, was perhaps the first to realise the potential they held for easing overcrowding. The key was that the largest use for electricity (whose provision had also been municipalised) was, in Edwardian times, still for lighting, so trams were run largely on cheap off-peak power during hours of daylight. Because local authorities could borrow relatively cheaply, trams could provide a dramatically less expensive way of travelling in the city. They were especially suitable for commuting to work over distances just a bit too long to walk; for such journeys the fare was one penny, and even a docker had to be a bit down on his luck to be unable to afford that sum.[36] The tram, quite simply, unlocked the overcrowding problem by allowing the middle classes to move out to new suburbs, freeing up older housing for the less well-off, and it allowed Brodie to provide temporary re-housing in Army huts while he cleared and rebuilt slum areas. Trams even earned a profit for the Corporation in the process.

When the *Town Planning Review* was launched in Liverpool in 1910 it was entirely excusable for most of its contributors to be fairly optimistic. Of course we now know that Corporation housing would never produce a complete answer because the backlog of need was so great that houses could not be built fast enough to keep up even with deterioration. Although we might pick on the great blocks of through-deck flats built in the 1920s as advanced for their time, by the time they were fifty years old they were mostly in need of replacement or extensive refurbishment, and in the worst areas housing of the type they had been intended to replace was still there. Liverpool's last "backhouses" were occupied well into the 1960s and, after standing derelict for over thirty years, are now being restored.

But this is by no means the whole story of the condition of the working classes in Liverpool. In the area to the south of the city centre, courts were less prevalent, though there were substantial numbers around Beaufort Street, and conditions in a relatively limited area just inland from the docks were truly appalling. As in the area below Everton Brow, crime and poverty fed off each other.[37] Happily, more typical of the South End were long rows of small but perfectly decent terraced houses, usually with about twelve-foot frontage and

---

1971).

[36]One pence represented about ten minutes' wages, but ten minutes at the present national minimum wage will not take a person anywhere at peak-hour prices.

[37]For an eyewitness account of the latter, see O'Mara, *Autobiography*.

about thirty-foot depth, including the rear outrigger with scullery below and bathroom above. Nearly all were "spec-built," often by those same Welsh builders who had evolved into property developers. The further south one went, the better the size and quality, the best of them boasting a pressed-brick front, a tiny "front garden" of about three feet and (perhaps the greatest status symbol of all) four chimney pots, indicating the presence of a fireplace in all the main rooms. These last were inhabited on the sometimes troubled frontier between the "aristocracy of labour" – highly skilled men earning perhaps fifty shillings a week – and mid-range clerks earning perhaps £100 or £110 per annum. "Per annum" sounds more respectable, but it is less money and you have to spend some of it on looking respectable.

The extent of fair-to-good quality terraced housing stretching out on three sides of the city suggests that we may have overestimated the severity of the workers' plight, and when we either look at the census figures or even apply just a modicum of common sense, we can see that the fact that Liverpool was dominated by transport and service industries did not necessarily make it an entirely low-skill economy. The operators of horse-drawn carts, wagons and delivery vans, for example, needed the skills of the shoeing smith regularly, as well as those of the saddler, wheelwright, wagon builder, blacksmith, coach painter and sign writer from time to time.[38] Ship repairing was an extensive and multi-skilled industry, as was the small amount of shipbuilding still going on. There were eight specialist firms of ship painters employing 490 people, and fifty-one paint manufacturers employed 586 people. These were heady days for the palm-oil trade – and hence profitable ones for a lot of coopers (2008 men, three women). The underestimation of mechanical engineering skills in Liverpool has already been noted, but one could go on: because ships were less seaworthy than they are now, it was not uncommon to arrive in port in need of the services of a boatbuilder to repair damaged lifeboats or even an anchor smith. All those clerks at their high desks generated work for 1106 bookbinders. There is no need to explore every side-turning, because the underestimates of skills rest on a simple failure to understand that skills exist where there is money to pay for them and that in Edwardian Liverpool there was plenty of money – as witnessed by the listing of over fifty coach and carriage builders in *Gore's Directory* for 1905.

For the physical evidence of that money we move a little further from the city centre, to Princes Park, where we find a wide variety of house types and styles, mostly built in the second half of the nineteenth century. The most modest of these are substantial three-story red-brick terraces (the third story was to advertise the fact that you kept live-in servants) with a road frontage of about

---

[38]In 1916, for example, 1223 men and eight women worked as wheelwrights. There were ninety-two shoeing smiths, and thirty-three sign writing firms employed an unknown number of people.

thirty feet, and between those and detached houses standing on half an acre or so it was possible to find something to suit most of the deeper pockets. Many hundreds of these houses still survive, mostly doing duty as flats for students and young couples. Further south again, the houses round Sefton Park are nearly all big enough to have their names marked on the twenty-five-inch OS map, but we are still only on the verge of the area where many of the seriously rich of Liverpool lived, a band stretching down from Woolton, through Calderstones, Allerton and Mossley Hill to the south end of Aigburth. Few of these pocket stately homes survive, most of their owners having sold out when you just could not get servants any more, leaving only traces like a road filled with 1920s semis having a common front wall in rusticated stone – the former boundary of some magnate's estate.[39] Others, as in so many cities, were turned into public parks.

\+ But we have seen enough to illustrate that the customary emphasis on a low-wage, low-skill economy, while containing some truth, does not give a balanced view. Certainly there was squalor and hardship, but there was much else beside.

One area of the stage on which our characters appeared remains to be mentioned: the sector providing services for shipping. It is little studied and poorly represented in archival material; indeed, much of it has left little trace except for entries in trade directories and what we may learn from the large OS maps. It has almost entirely disappeared now, swept away by the smaller number of ships which turn around faster and have much smaller crews. In the days when a run-of-the-mill cargo liner had a crew of eighty or so and cruised at about ten knots, a lot more eating got done on board, making ship provisioning a decent living for many small firms: the provisions for an express liner with 2000 or so passengers (plus several hundred crew) aboard for five or six days were another matter entirely.

Keeping ships looking smart used to be difficult as well, especially again in the case of passenger liners. An odd copy of a Cunard stores book which has survived (unfortunately from a later date) gives some idea of the potential market these ships represented for local suppliers.[40] In 1935 *Aquitania* purchased 636 tins of bath powder, 972 tins of metal polish, 409 kegs of soft soap, 760 gallons of liquid soap, 492 scrubbing brushes, 294 carpet brooms and much more. But this was only for the boot stores: the deck stores used, among much else, nearly 1000 gallons of paint and varnish, 640 gallons of liquid soap and fifteen dozen brooms. There were, of course, other stores, and when we are reading of various liquid soaps and detergent literally by the thousand gallons we must remember that this was an oil-fired ship – a coal-fired one of 1905 would take much more to keep it

[39]Happily, the Holt family's Sudley House is preserved by the NML.

[40]NML, MAL, B/CUN/4/12, Cunard stores book, 1935.

clean. In the 1900 trade directory for Liverpool there are eighty-five ship's store dealers and nearly 1000 entries which begin with the word "ship." There were, of course, other businesses not involving the word "ship," such as five lifebelt makers, and one presumes that the employment of 2530 people in oil, grease and tallow refining was not unrelated to the replenishment of ships' engineers' stores.

It is a simple matter to estimate that when a large liner arrived in Liverpool with a more-or-less full complement it would land something of the order of 40,000 items of laundry, over and above that done on board. (The first-class table napkins alone, if every passenger took three meals a day, would came to 8000-9000) This is why 364 Liverpool laundries employed nearly 4000 people.

The city in which the continued expansion of the port took place was therefore a place of violent contrasts, with a local economy was more complex and varied than has usually been acknowledged. This is not to deny that it was very much a port-based economy but only to point out the extent of that base.

Leisure in Liverpool

Since at least the end of the eighteenth century, Liverpool had considered itself the "Athens of the North." Class considerations dictated that there were cultural facilities for "us," such as the Philharmonic Hall, and for "them," such as St George's Hall. For gentlemen there were the libraries of the Lyceum or the Athenaeum clubs, for others there was the Public Library. But before we go too far in mocking such paternalist provision we should remember that municipal support for cultural leisure activities was of a high standard and one did not need much money to become familiar with great art, literature, music or drama. For the poorly but regularly paid it was more affordable in 1905 than it is now, particularly if one takes account of the tram fare for getting to the Walker Art Gallery or other "provision." The Central Library was able to offer better opening hours for those aiming to improve themselves by attending evening classes than it can now.

Because so much of the waterfront was cut off from the town, there was not much opportunity for people of any social class to stroll about watching the ships coming and going. Since almost everyone in Liverpool had contact with the sea either directly or at a single remove, this opened a considerable leisure market for the various companies operating ferries, particularly those going upstream to Eastham or downstream to New Brighton. The latter was a full-blown seaside resort, with (as they kept telling everybody) a tower higher than Blackpool's, whose grounds offered a variety of amusements including an American roller rink, a bazaar, menagerie and such white-knuckle rides as a water chute and a Himalaya Railway. Proud boasts were made as to the quality of the sandy beach, though these relied heavily on the public's ignorance of the sewage disposal systems of Merseyside. For the better-heeled there was the possibility of a day-trip to northwest Wales or even as far as the Isle of Man.[41]

Another leisure provision which was plentiful in Liverpool was large public parks in which to stroll. There has been some confusion over their origins, with them being portrayed either as a wonderful paternalist benefit or as a hypocritical way for the upper commercial classes to hog a lot of land. In reality they were both and neither, for they relied on one of the classic nineteenth-century financial conjuring tricks. Take a large patch of land, divide its perimeter into building plots of generous size. Lay out the middle in a combination of carriage drives, rides, paths and open spaces and provide one or two "attractions" like a lake or a palm house. Dedicate these to "the public." Now sell the building plots. They have a pleasant and unobstructed view in perpetuity, which so enhances their value that the proceeds pay most or all of the costs of laying out the park. Everybody gains: Joe Public has a pleasant place to go; the Corporation gets some good PR and its money back; and the rich folk are spared the burden of having to buy and maintain sufficient land that they cannot see another building from their front windows. A trick as good as that was worth working repeatedly - at Prince's, Newsham and Sefton Parks.

There were, of course, darker sides to leisure in Liverpool. Prostitution was endemic, driven on the one hand by need or greed and on the other by short-term "riches" and lust. In that respect Liverpool was little different from other seaports. Unfortunately, another of its favourite leisure activities, shared with few other ports, was the matter of religious "differences." In fact the parties were in complete agreement on the key point, which was that religion formed a convenient way of picking two teams to have a really good fight. The levels of provocation applied by both sides were extraordinary, and so long-lasting was the hatred and suspicion generated that its last traces were still recognisable into the late 1970s, when the remarkable partnership of Bishop Shepherd (Anglican) and Archbishop Worlock (Roman Catholic) finally laid them to rest.

Vulgar entertainment is not very well-documented; nobody has done for the Edwardian years what Hugh Shimmin did for the mid-Victorian period.[42] It seems clear from the continued survival of truly huge numbers of pubs, bars, beerhouses and the like that drink continued to play a key role in both entertainment and impoverishment. O'Mara and others testify that drinking beyond one's

---

[41]It is interesting to note that in 1904 the journey time was given as 3.25-3.5 hours. That is faster than today's "boats," although not as fast as the Seacat.

[42]See J.K. Walton and A. Wilcox (eds.), *Low Life and Moral Improvement in Mid-Victorian Liverpool. Liverpool through the Journalism of Hugh Shimmin* (Leicester, 1991), for an account of Shimmin, a useful selection of his writings and a helpful introduction.

income was routine among "slummies," resulting, as elsewhere, in aggravated deprivation and the wife-beating that went with arguing about money. But we also see the emergence of what were rudely known as gin palaces, which were in fact aimed at a much more respectable clientele than were the earlier recipients of that title. Two particularly fine *art nouveau* cxamplcs built for brewer Robert Cain survive in Liverpool, but however successful they may have been as individual ventures, drinking in pubs in the evening had yet to become respectable.

Lunchtime drinking (and Edwardian lunchtimes could be very elastic) had two roles. It was in part simply a matter of leisure, an impression which was later confirmed by Lloyd George's strictures (and legislation) on the harmful effects of working-class drinking. But this impression conceals the fact that in an economy which depended very heavily on casual work and on the ability to find a gang of skilled men, ship painters perhaps, at short notice, the pubs were an essential networking tool.[43] This was in some senses corrupt, in that those seeking work clearly felt some pressure to buy drinks for those offering it, but it was the most effective system available at the time. The hinge point was close, though, with the rapid spread of the telephone affecting the old system of "business in person" at all levels of the local economy.[44]

It was a hinge point, too, for those whose leisure was involuntary. The gradual improvement in working conditions for those who had work and the appearance of the first rudiments of city-wide organisation with such initiatives as the Liverpool Council for Social Service for those still less fortunate mark a fairly clear break with a paternalist past. There were two aspects to this: the old style of gentleman who gave not only of his money but also of his time to a variety of local charities had been well-intentioned but disorganised so that the results were necessarily patchy. The anecdotes of shoeless waifs managing to log four or five Christmas dinners are probably apocryphal, but the "Official" section of *Gore's Directory* for 1905 does have ten pages of entries for charitable institutions. Clearly a bit of coordination would do no harm. At the same time there was a visible tendency towards supporting other "good causes," particularly the University, which proved to be a pioneering institution in the development of

---

[43]Casualism was definitely not confined to dock labourers but also was widespread in ship repairing and other trades.

[44]This is neatly illustrated in Liverpool Record Office, Liverpool Overhead Railway Papers, 1914 General Meeting, minutes, where the increased use of telephones is given as a reason for falling passenger numbers. Conversely, in rural areas where the telephone penetrated only slowly, pubs continued to be allowed to remain open all day on market days.

both town planning and social services.[45] But almost opposite the University on Brownlow Hill and the inappropriately named Mount Pleasant still loomed the ugly bulk of a workhouse rebuilt and enlarged in 1850 to contain 3500 inmates. It was widely hoped that the former would help stimulate a new skills-based local economy, thus removing the need for the latter, but the roots of "institutionalised poverty" were to prove very deep.

To the general question "what was it like to live and work in Liverpool in 1905," one can only give the same answer that is appropriate in almost any time or place: it depended upon who you were.

[45]See M. Simey, *The Disinherited Society. A Personal View of Social Responsibility in Liverpool during the Twentieth Century* (Liverpool, 1996).

Chapter 1
The Port and Its Trade in 1905

At first sight, then, the physical condition of the Port of Liverpool in 1905 was fairly impressive. Thanks to the new (1901) Sandon Entrance, Canada and Huskisson docks could handle the largest ships in the world and would suffice for such famous ships of the immediate future as *Mauretania* and *Lusitania*. The engineering works at Kings and Queens was another success story: they were large and spacious docks but, dating from the late eighteenth century, were shallow and actually the wrong shape for steamships, which needed both wider quays and a higher quay length/water area ratio. Queens had undergone a fairly successful facelift in the early 1890s which more than doubled its typical annual dues income.[1] But the new scheme was much more radical: Kings and Queens became "spine and branch docks." In the case of Queens, its water area and linear quayage increased by some fifty percent but its earnings increased from £19.67 per yard per year in 1899 to £44.47/yard in 1909 (the first full accounting year after completion of the whole scheme).[2] There was a long and fairly steady rise of the port's total dues income between 1900 and 1907, but it totalled only about twenty percent, making Queens a far above-average earner and at first sight an impressive investment. In fact, of course, what further investigation shows, as common sense would suggest, is that trade at Queens could only grow at a well above-average rate within the overall growth by "stealing" trade from the below-average docks within the port. If we were to assume that every last penny of the £204,318 increase in total dues income between 1900 and 1909 was directly and solely attributable to the improvements in dock facilities under the Acts of 1891 and 1898, which represented the investment of over £5,000,000, then the marginal revenue would represent just under three percent on capital, noticeably below the rate at which the Board had borrowed the money to pay for the works.

---

[1]Income rose from around £13,000 before the works to around £29,000 afterwards. These and similar figures below are taken from the sequence National Museums Liverpool (NML), Maritime Archives and Library (MAL), Mersey Docks and Harbour Board Records (MD&HB), Revenue Statements, 1844-1924. The "facelift" involved increasing the water depth by impounding pumps, broader quay margins on the east side and a new shed.

[2]There are some doubts about the best way to express the efficiency of docks: revenue and/or tonnage per linear yard of quay were used at the time and have generally proved popular with historians of such things. For reservations about "efficiency" calculations, see appendix 1.

These docks looked good and worked well, but their financial performance was less impressive. Yet again we are reminded of John Bramley-Moore and his statement that it was impossible to operate docks at a profit.

Just as important was to provide better river entrances at both extremities of thc estate and to combine the docks into a coherent system. The programme was enabled by the 1898 Dock Act with an estimated cost of over £3,500,000. The confidence in the success of these works exhibited by members of the Mersey Docks & Harbour Board (MD&HB) and its customers was to be justified by the repetition of figures like those given for Queens once the whole scheme was completed in 1908.[3]

Figure 2
Part of New South Works under 1898 Act

Source: National Museums Liverpool.

[3]But this did not occur immediately; see below.

But other docks were in a much less satisfactory state. The money for modernisation would not stretch to all parts of the estate, and in 1905, Canning earned just £2317 in dues on goods and ship tonnage together, of which £1299 was on ship tonnage, almost all (£1260) of which was paid by coastwise vessels. This is partly a reflection of the lower dues on coastwise goods and partly of the fact that Canning was home to Liverpool's small but quite effective fishing fleet. Despite the engineering works which were temporarily forcing ships back into some of the older, smaller docks, these figures compare unfavourably with those for 1848 (the first full accounting year for which the Canning system could be considered complete), at £3013 total, £2563 of which was on tonnage.

They are unfavourable figures in another way too. Those docks handling the port's chief traffic generally earned over half of their total income from dues on goods.[4] The fact that Canning's proportion is well under that indicates that some of its apparent bustle was in fact the result of vessels laying up or entering only to use the graving docks. None of the vessels, of course, were large, for the entrance and passages of the system varied between forty-four feet three inches and forty-four feet nine inches in width and the depth on the entrance sill at the top of mean high water was about twenty-one feet. The figure on which Canning stands finally and totally condemned, however, is its earnings per linear yard of quay, which stood at £2.28 - a considerable improvement on 1890, when it earned £1.20 per yard. In 1905, Hornby, the newest dock in the system, earned £62.94 per linear yard.

Albert Dock handled little shipping even during its Indian Summer brought about by the non-availability of other berths, but it did better than Canning, at £12.97 per yard.[5] But its total income from dues, at £11,477, was well below that from its huge warehouse blocks, which were busy with transfer cargoes carried by road, rail or barge to or from other docks, and earned a net profit to the Board of £17,050. It had, however, long been recognised as obsolete, and schemes for the global modernisation of this whole sub-system of the docks had already been considered.[6] They would continue to be considered,

---

[4]Hornby, which opened in 1896, was for a time the premier dock in the high-value US cargo trades and earned £91,959 in total dues, of which £47,348 was on goods. Alexandra, a much larger dock, earned £141,199 in total and £75,433 on goods.

[5]For an appraisal of the efficiency of Albert, see G. Milne and G. Tonks, "Specialised Port Facilities on Trial: Liverpool's Albert Dock in the Nineteenth Century," in A. Jarvis and K. Smith (eds.), *Albert Dock: Trade and Technology* (Liverpool, 1999).

[6]For further details, see M. Stammers, 'A Dock Where Nothing Will Ever Be Done: Albert Dock in the Twentieth Century," in Jarvis and Smith (eds.), *Albert Dock*.

as in the somewhat farcical "Great Fish Dock" scheme of the late 1920s.[7] But in 1972, when the decision was taken to close all the docks upstream of Pier Head, this group remained almost exactly as Jesse Hartley left them in 1847, which was to be good news for the Merseyside Maritime Museum.

This seems a quite extraordinary policy for the Board have pursued. They were perfectly well aware of the notion of the "pet dock," and Canning and Salthouse were pet docks, which is why the shipping tonnage per yard of quay using them, despite their crippling disadvantages, was low rather than pathetic.[8] Stuart Mountfield has drawn attention to the debate which occurred within the Board between "upstreamers" and "downstreamers."[9] There also was a shortage of members who took seriously the modernisation of docks on what was potentially the port's most desirable site.[10] The completion of the new Dock Offices in 1907 would make it even more desirable, but it was already within a few minutes' walk of all the Board's departmental offices except the Engineer's, as well as the Custom House, and the area was thick with suppliers of every kind of marine stores. It was precisely this sort of location that has caused Gordon Jackson on a number of occasions to note that in the final analysis a port is not a place, but a community of merchants.

In short, if the success of the port were to be measured by the size of the investment made, it would be adjudged overall as highly successful. But matters were not so simple. As Jackson has shown, increased revenue could lag considerably behind investment, with unpleasant effects on the finances of port authorities.[11] The next stage therefore is to consider the state of Liverpool's finances in 1905.

The central fact is that Liverpool, like virtually every other port authority, had enormous debts, to the extent of £24.4 million in 1905, up from £22.5 in 1904, and the Board held unexpended borrowing powers for a further

---

[7]NML, MAL, MDHB, Newscuttings, III/40. For further details of the Liverpool fishing industry, see A. Jarvis, "An Historical Backwater: The Fishing and Fish-trading of Liverpool," *Northern Seas Yearbook* (1995), 51-75.

[8]Canning's figure was 85.42 tons, while Salthouse stood at 64.44.

[9]For further details of this long-running disagreement, see S. Mountfield, *Western Gateway: A History of the Mersey Docks and Harbour Board* (Liverpool, 1965), and below.

[10]The reader might assume that the Pluckington Bank would stymie such proposals: see chapter six for a demonstration of the overconfidence the Board exhibited in its dredging capabilities.

[11]G. Jackson, *The History and Archaeology of Ports* (Tadworth, 1983), 117.

£2.2 million.[12] This was reflected in interest costs that year of £833,495, or forty-nine percent of the Board's total revenue. Because nearly all the money was borrowed using bonds, rolling re-financing (i.e., redeeming expired bonds and issuing new ones) was always undertaken to a greater or lesser extent, which meant that the organisation's finances were highly vulnerable to rises in prevailing rates of interest.[13] A rise of one percent in the average rate of interest on the Board's debt would cost £244,132 per year, significantly more than the total revenue cost of running both the Engineer's and the Marine Surveyor's departments.[14]

In this respect, the Board had enjoyed a long run of good luck with low interest rates: between 1870 and 1900 their debt had increased from £14.4 to £23.9 million, but their interest costs had risen only from £642,509 to £679,356, which represented a significant and continuing decline in the percentage of their dues income spent on interest from a fairly terrifying eighty-six percent in 1870 to a sustainable sixty-two percent in 1900. By 1905 the writing was on the wall: interest costs had risen to £833,495, or sixty-five percent, by 1907 to sixty-seven percent and by 1909 to seventy percent of revenues. Even in 1911, when a new record tonnage used the port, interest payments slipped only to sixty-nine percent of revenues. The reason is easy to find: old bonds at interest rates as low as 2.5 percent were slipping off the bottom of the balance sheet as they were redeemed while new bonds were being issued at 3.75 percent. Even the new 1905 debenture stock, on which so much hope had been placed as a cheaper way of financing the constant investment in new works, was paying 3.5 percent. There was every danger that over a period of time the notional one percent rise mentioned above would become a reality, in which case the condition of the Board's finances would become dire indeed.

Continuing to invest in one major programme after another against that financial background might be seen as more foolhardy than courageous. There are two strands to the question.

We may observe a number of long-term trends, of which the most basic was the growth of the throughput and revenue of the Port of Liverpool. This had been so long-standing and consistent that it is entirely understandable that the

---

[12]These and all similar figures are taken from the MDHB's published *Annual Accounts*.

[13]Liverpool Dock Bonds were a long-standing form of finance and were effectively a mortgage on future dues income, on the basis of which they were secured. They were *not* secured on the value of the Estate.

[14]Though, of course, only a comparatively small proportion of bonds came up for renewal in any one year, so such a rise could only occur over a period of time.

Edwardian gentlemen might have expected it to be irreversible. The tonnage of shipping using the port first exceeded one million in 1823, two million in 1838, three million in 1845, four million in 1854, five million in 1866, six million in 1871, seven million in 1877, eight million in 1882, nine million in 1888, ten million in 1895, and an amazing 15,626,241 in 1904.[15] There were good years and bad: the speculative boom and ensuing slump of 1824-1825 produced a dip in revenue, though not in tonnage, and the 1847-1848 commercial crisis was reflected in both figures, but growth soon resumed. Even the American Civil War produced only a minor dip, which was over by 1866. As will be seen from the figures above, the time required for tonnage to grow by one million declined at an increasing rate.

The so-called "Great Depression" of the late nineteenth century did not even achieve the status of a myth in Liverpool shipping circles.[16] In such a complex thing as shipping one can, of course, produce figures to support any opinion, and in this case it is only necessary to draw a graph of the year-on-year growth of tonnage using the port to come to a very different conclusion: over the period 1850-1900 it shows, amid violent fluctuations, an unmistakable downward trend, with each peak lower than the last and each trough deeper. This, however, was a view taken only by a small and pessimistic minority: to the majority, trade had always grown and was still growing at an accelerating rate. Just as their predecessors had "toughed it out" on previous occasions when outside factors made life difficult, so they would now.[17]

The great diversity of the port's trade was another comforting factor, because over the long term it was extremely unusual for a variety of trades to suffer bad years together; indeed, in some cases it is arguable that a bad year in one trade could cause a good year in another. The extreme example is that when the potato import trade had a bad year, the emigrant trade might well have a good one, as happened notoriously at the time of the Irish Potato Famine, but a mini-slump in 1907-1908 sufficed to cause a brief upturn in the number of emigrants.[18]

---

[15]Tonnage alone, however, can be a misleading measurement of trade: in 1902 the aggregate ship tonnage rose from 12.6 to 13.3 million, while the dock rates on goods fell from £347,898 to £344,091, indicating more ships sailing partly or wholly in ballast.

[16]Cf. S.B. Saul, *The Myth of the Great Depression, 1873-1896* (London, 1969).

[17]For the extraordinary story of how the old body of Dock Trustees deliberately spent counter-cyclically during the 1847-1848 crisis and was able to carry on borrowing when the Bank of England could not, see A. Jarvis, "Port History: Some Thoughts on Where it Came from and Where it Might be Going," in L.R. Fischer and A. Jarvis (eds.), *Harbours and Havens: Essays in Honour of Gordon Jackson* (St. John's, 1999).

[18]Reported in *Engineering*, 14 February 1908.

The reader may wonder whether the average Board member had the information readily available to enable him to take that kind of long view: he did, for every issue of the *Annual Accounts* included basic statistics of the port's trade for every year since 1752.

The self-belief of the Board could be astonishing. In 1875, when the capital expenditure on the works of the 1873 programme was getting up to speed, the Board received an offer of £500,000 on bond from the Lords of the Admiralty as part of the endowment fund for Greenwich Hospital.[19] The negotiations which followed show that the Admiralty understandably thought the Board would be applying for a loan, while the Board was accustomed to people applying to it for bonds. Rather than compromise that position for a mere £500,000, the Board stalled the negotiations, though given the absurd positions adopted by the Admiralty that was not difficult. Such a policy was not the folly it sounds: at an early stage the Board's solicitor had advised the Chairman of the Finance Committee that "we should run the risk, however remote, of having the present good credit of the Board damaged by the fact becoming known that a Government Department, after investigation, had declined to lend on our Securities." The Board was well aware that its creditworthiness was its critical asset on which all else hung: the problem is that by 1905 the world had changed, and the failed debenture issue of 1900 had ended forever the days when it could pick and choose its lenders. Not all the members understood this, and some thought that the buoyant tonnage figures were sufficient to sustain previous levels of self-belief.

The second long-term trend which was not merely observable but which positively thrust itself in members' faces was the increase in ship size, which can be viewed in two ways: the growth of the typical ship or of the exceptional one. Liverpool's policy had long been that if it was to maintain its place in the world it must continue to increase the size of some of the docks and their entrances so that the largest vessels in the world could use them. The first wave of this development occurred between 1840 and 1860, when the width of the largest entrance in the port increased in stages from about forty-five to 100 feet. As each "biggest" dock was superseded, the exceptional vessels moved on to the latest dock, leaving their former home to lesser craft. Of the docks in use in 1905 only a handful had been built for vessels of the class then using them; the majority had started out in the North Atlantic trade, long the user of the largest and most demanding vessels.[20] There were thus two factors to consider in deciding whether or not to build a new state-of-the-art dock for the exceptional vessels: whether its

---

[19]The documents for this episode are gathered in NML, MAL, Unbound Worked-up Papers (UWUP), B122.

[20]For example, Collingwood (1848) was the last new dock actually designed for the coasting trade before the improvements described in chapter three.

use by those vessels would generate sufficient revenue and whether there were lesser vessels waiting to step in and use the Canada-Huskisson system which, in 1905, had the deepest berths in the port. There was a good reason for confidence: during the years 1897-1905 the size of an average ship in the Liverpool-Boston trade rose from 5971 to 12,213 tons, and there was every reason to believe that growth in the size of cargo and passenger vessels was accelerating and would continue to do so.

In 1904 Sir William White, a most distinguished naval architect (Director of Naval Construction, 1885-1902), was installed as President of the Institution of Civil Engineers, and his Presidential Address is a very substantial summary of the state of naval architecture, covering warships and merchant ships alike.[21] It must have been chilling stuff for any dock engineer present, as he outlined the sizes of ships of the 1860s, typically little over 1000 tons and drawing around twenty feet or less, and then moved on, via a consideration of *Great Eastern*, to explain how "a first class port should have provision for ships up to 1,000ft long, with 100ft breadth in entrances and 33-35ft depth of water." He gave figures which showed that the average gross register tonnage of all vessels registered at Lloyds reached 1628 by 1902, and that of all steamers 1781. Excluding small vessels of under 100 tons raised the latter figure to 2200. "Ten years ago, there were few vessels of over 10,000 tons. At the end of 1902 there were 84 steel steamers of 10,000-20,000 tons." Worse, he drove home the message, with a string of examples, that the largest ships afloat could safely be loaded to a much greater depth than was presently possible due to lack of depth in ports, resulting in an increase of coal and cargo capacity of roughly one third.

The link between size and speed in designing large passenger liners is well known, but there is a less obvious factor in the background which required that each new step taken in designing for the North Atlantic had to be bigger than the last. It was no use gaining a few hours on the voyage time: the need to arrive at high tide meant it was twelve hours or nothing. That is why Sir William correctly forecast that the next step up from Norddeutscher Lloyd's *Kaiser Willhelm II* would have to exceed its size by a much greater margin than that by which it had exceeded that of *Campania*, which previously held the Blue Riband. Even he, however, was wrong about draughts, for the thirty-three to thirty-five feet he suggested as adequate for a first-class port was within three years shown to be inadequate by the very type of ship to whose building he had referred, *Lusitania* and *Mauretania*, which drew over thirty-six feet fully laden.

On balance, the evidence seems to be that the Board members who advocated caution on the grounds that most of the port's dues came from average-sized ships and that only the North Atlantic ships made exceptional demands on

---

[21]W.H. White, "Presidential Address," *Minutes of the Proceedings of the Institution of Civil Engineers*, CLV (1903-1904), 2-169.

ports were wrong.[22] It was not only the exceptional ships which were growing but the average ones as well, and the efficiency with which those average ones operated was fairly closely connected to their draft. Here, for once, was a case of the spectre of foreign competition, so often the first resort of scoundrels, posing a genuine, or at least credible, threat. This was the time at which Greece began its remarkable rise to prominence in shipowning, and the perceived threat posed by German merchant shipping is well known.[23] Nor was it just a matter of speed: the German Atlantic liners consistently loaded more passengers per voyage than any of the services originating in Britain.[24] If Britain were to retain its place as the leading shipowning nation, it had to have the best ships, and to do that it needed ports with more depth of water.

There had been a time when it was possible to claim that there was no point building bigger docks in Liverpool because the facilities at the other end of the port's main trades were so poor that only smallish vessels could be used. By 1905 that argument was only barely tenable, as extensive dock improvements took place in ports overseas. Within ten years, ports on many of Liverpool's main trading routes either had made or were making improvements to provide depths of thirty feet or more.[25] On the other hand, in 1905 Sandon was the only

---

[22]This division was not peculiar to Liverpool: Sir William Mathews, in his Presidential Address to the Institution of Civil Engineers, mentioned the unwillingness of the owners of average-sized ships to face possible increases in port dues made necessary by investment needed only for the exceptional vessels as a factor inhibiting port development, which was in turn inhibiting progress in naval architecture. *Minutes of the Proceedings of the Institution of Civil Engineers,* CLXXI (1907-1908), 5-48.

[23]For Greece, see G. Harlaftis, *A History of Greek-owned Shipping. The Making of an International Tramp Fleet, 1830 to the Present* (London, 1996), figure 4.1 and table 4.10. By 1902, for example, Germany owned nearly 2.8 millions tons of steam shipping (compared with Britain's 13.4 million); what was more worrying was that of eighty-four ships in the world between 10,000 and 20,000 tons, Germany owned twenty-six against Britain's thirty-nine. White, "Presidential Address."

[24]*Engineering* used to publish very thorough reviews of the state of the shipping and shipbuilding industries, including passenger figures for the major lines, spread over two issues each January. Figures to support the above statement occur, for example, in the issue of 20 January 1905, 91. It should be noted, however, that Cunard's Fiume-New York service outperformed the two German lines.

[25]Great Britain, Parliament, *Parliamentary Papers* (*BPP*), 1917-1918, IX, 523 (Cd 8641), Royal Commission on the Natural Resources, Trade and Legislation of Certain Portions of His Majesty's Dominions, contains "Memoranda and Tables as to Harbours and as to the Panama and Suez Canals" which detail improvements in most of the major ports of the British Empire. Institution of Civil Engineers Archive, London, Original

Liverpool river entrance which could provide such a depth on a regular basis.[26] But the new Brunswick entrances under construction would in theory give over thirty feet at mean high water.

The Board also had the problem that the companies which owned the biggest and deepest ships - principally Cunard and White Star - did not own just a handful of huge ships, and there was the risk that if the traffic of the express passenger liners were lost to another port through failure to keep abreast of their ever-increasing size, then that of the smaller vessels of the same companies, serving many routes as well as the North Atlantic, would be lost as well. Moreover, these were exactly the class of ship which provided much of the port's bread and butter. In the event, when first White Star and later Cunard moved their express liners to Southampton, the "slow boats" did not all follow, and the effect on the port's revenue was negligible. But nobody knew that in 1905.[27]

Out in the wider world was another long-term trend: the growth of oceanic trade. There might be good years and bad, but overall growth had been continuous for so long that once again the gentlemen of the Board might have been forgiven for thinking it irreversible. For example, the total UK tonnage of vessels in foreign trade entering British ports grew from 20.4 million tons in 1871 to 60.1 million tons in 1906.[28] Again we must be cautious in relying too much on tonnage figures, for in the recession year of 1892 the number of coastwise vessels entering "with cargo only" rose, despite the fact that their aggregate tonnage fell. The tonnage of entries "with cargo and in ballast" rose considerably, suggesting that the larger coastwise vessels were having a hard time getting full cargoes while the smaller ones fared relatively better. A similar situation occurred with foreign trade vessels in 1890, when the number entering with cargo fell by less than one percent but their tonnage plunged by sixteen percent.[29] This conclusion would not be possible from the tonnage figures alone. Such considerations were absent when simple tonnage figures were used to "talk up" the prospects of the port, but since the overall growth was large and steady it did not really matter:

Communications No. 3618, L.F. Vernon-Harcourt, "The Milan Navigation Congress of 1905," gives brief details of improvements in Genoa, Venice and Spezia.

[26]The reason for the stipulation "regularly" is that the difference between average high water of neap tides and of springs was more than sixteen feet. On a large tide, Herculaneum could just about manage thirty feet.

[27]See chapter five.

[28]D.J. Starkey (ed.), *Shipping Movements in the Ports of the United Kingdom 1871-1913: A Statistical Profile* (Exeter, 1999), 73.

[29]*Ibid.*, 354-355.

the perception, as with the growth in ship sizes, was that growth in trade would mean that new docks were always going to be needed.

Nor, as Jackson has pointed out, need we worry much at this stage about the decline in Liverpool's share of the total port services industry, to which one or two pessimists at the time drew attention and which was brought to the notice of historians by Francis Hyde.[30] While Liverpool's share of vessels entering from foreign ports fell from 19.5 percent in 1870-1874 to 13.9 percent in 1900-1904, this relative decline largely reflected the growth of the coal export trade, particularly through the South Wales ports, where heavy investment in deep-water docks with huge entrance locks and in coal-handling machinery racked up enormous tonnages.[31] The fact remained that Liverpool and London (which was also in "relative decline" when measured by tonnage) dominated when their trade was measured by value rather than volume, with 26.4 and 29.3 percent, respectively.[32] The overall tonnage figures above form a perfectly adequate basis for optimism as to the long term, and members of the Board were well used to thinking that way because dock works took so long to design and construct. Indeed, in dock development there was usually no such thing as a "short term."

On the other hand, 1905 was generally a worrying year. In 1900 freight rates suffered their sharpest fall in a decade, with outward rates plummeting by half, but by 1905 they were climbing sharply again as observers would have hoped and expected. The problem was with homebound freights, which recovered about a quarter of what they had lost before starting to dip again in 1904.[33] This, perhaps surprisingly, did not cause any marked slackening in the growth of the Board's income, though using the slightly more sensitive indicator of dues on goods alone (i.e., excluding tonnage dues) there was perhaps cause for some concern. Looking inland may well have given cause for even more.

As it did for the shipping industry, *Engineering* published an overview of the previous year's trade and industry each January, and in 1905 the view of the previous year was not pleasant: it was characterised as the most depressed year in the past eighteen. "Such a lack of work is happily infrequent, but returns from time to time" was the only consolation to be offered in "diminishing

---

[30]This theme runs throughout F.E. Hyde, *Liverpool and the Mersey: An Economic History of a Port, 1700-1970* (Newton Abbot, 1971).

[31]These were not required to accommodate very large vessels but a multitude of smaller ones because of the huge tidal ranges found along the South Wales coast.

[32]Jackson, *History and Archaeology*, 113.

[33]A.W. Kirkaldy, *British Shipping: Its History, Organisation and Importance* (London, 1914; reprint, Newton Abbot, 1970), appendix XVI.

employment, culminating in dire distress." The engineering trades were not doing too badly, but the cotton industry was "bad;" building, woodworking and furniture trades were "slack;" and the boot and shoe and clothing industries were "depressed."[34] Cotton, timber and hides were, of course, imported through Liverpool in significant quantities. Nationwide, unemployment of trade union members stood at 7.6 percent, with the main groups suffering above-average rates including plumbers (ten percent) woodworking and finishing trades (10.5 percent), carpenters and joiners (11.5 percent) and workers in the leather trades (11.9 percent).[35] The cotton and wool trades were said to be showing signs of revival.

Rather closer to the hearts of many of the members of the Board was the fact that shipbuilding was definitely recovering, indicating enhanced optimism in the shipping industry. In 1904 the recovery in tonnage built had begun, though it was said that many contracts had been agreed on unremunerative terms to keep the yards working, and output on the Clyde was still falling. In 1905, however, there were significant rises in every significant centre except the Mersey, most notably on the Clyde, where tonnage under construction rose from 432,400 in 1904 to 540,000 in 1905.[36]

At first glance, the Board seemed therefore to be in an excellent position. The tonnage of shipping paying dues to the Board was continuing to rise; indeed, it rose by nearly a quarter between 1900 and 1905. Within that total, some of the rise was in tonnage paying harbour dues only, i.e., vessels using other ports on the Mersey, principally those on the Manchester Ship Canal, but the greater part - over 2.75 million tons - was in vessels using the docks. Moreover, in terms of market share Liverpool seemed to have turned the corner: having suffered a small but steady decline in market share throughout the second half of the nineteenth century, its share of the recent national recovery in traffic was large. Between 1899 and 1903 the total increase in value of British imports and exports was 10.8 percent, or in money terms £88.4 million. Of that, Liverpool accounted for £36.1 million, or nearly forty-one percent. That share of growth was greater than those of London, Manchester, Glasgow, Bristol and Hull combined. The fact that revenues had not rocketed was simply due to the fact that the Board was a non-profit body, and when it was prospering it could and did cut rates to its customers: if 1904's traffic been charged 1884 prices, it would have placed another £427,000 in the Board's coffers.

---

[34]*Engineering,* 6 January 1905, 27.

[35]*Ibid.*, 27 January 1905, 131.

[36]*Ibid.*, 6 January 1905, 12-13; and 5 January 1906, 13.

The opening of the Manchester Ship Canal had made minor inroads into Liverpool's staple traffic of cotton, but the Mancunian failure to develop the essential mercantile skills which were traditionally plied in Liverpool meant that much of its traffic at this date was in comparatively low-value goods like timber, some of which was "new" business, and much of what was not had been stolen from Fleetwood.[37] The impact on Liverpool was not great: averaged for each of the five years to 1903, Liverpool had handled between seventy-four and seventy-eight percent of all British cotton imports, but in 1903 this had leaped to eighty-five percent. Another staple, tobacco, was showing a gratifying return on investment in the Stanley Tobacco Warehouse, and a record total of 125,000 hogsheads was earning rent for the Board in 1904. A more homespun indicator of the state of trade was also favourable, namely that Liverpool supplied more bunker coal to foreign-bound steamers than any other port; after all, the more fuel that got burned the more work was being done. Much of that coal was, of course, being burned to enable passenger liners to go at twenty knots rather than nineteen, making it a somewhat erratic indicator.

The verdict on whether and when it would be prudent for the MD&HB to venture on building its largest dock to date was particularly problematic, and the Board spent a good deal of time trying to foresee the future.[38] The attitude which prevailed in 1904 when it first authorised site preparation works for Gladstone, and the 1905 decision to go to Parliament seems to have been based on the old philosophy, which dated back at least to the 1840s, that by the time the works were complete, trade would have picked up again and the new facilities would be needed urgently. The realisation that the interest problem was not going away and that a high proportion of the bonds were coming up for re-financing prevented the Board from commencing serious works under the 1906 Act until 1908, when a sudden attack of cold feet occurred after a few months and the works were suspended.[39] This concern was well justified, for the Board was actually on the brink of insolvency, and, as emerges from the Admiralty episode above, the members knew that if they jeopardised their good name as a borrower the consequences could well include the complete collapse of the port because the inability to borrow at an affordable price crippled the ability to modernise the docks and their equipment.

---

[37]Fleetwood lost so much trade to Manchester, especially in timber, that only the famously rapid growth of steam trawling kept the port alive.

[38]There is more on this in A. Jarvis, *Liverpool Central Docks 1799-1905. An Illustrated History* (Stroud, 1991), chapter nine.

[39]For the detailed chronology, see chapter two.

Nor were the problems of 1908 a flash in the pan: in 1905 the Board had new borrowing of £3,027,384 in 3.5 percent debenture stock and £908,818 in 3.75 percent bonds. In 1906 the Board issued another £1,002,110 in debentures, of which £648,770 involved the refinancing of old bonds at lower rates of interest. In 1907 another £1,215,639 of debentures were issued, along with £661,637 in bonds at four percent; of the total, £1,643,002 was refinancing, leaving just £234,274 for the continuation of the works under the 1898 Act, and there was no improvement in 1908, with another £1,831,500 of four percent bonds refinancing £1,086,271 of old ones, virtually all of them at lower rates of interest. Although the Board had long realised that bonds were an expensive form of finance, the suddenness of the turnaround to being a serious albatross seems to have taken many of the members by surprise. In the long term, this huge and costly borrowing was to be a major factor in the collapse of the Board and its replacement by the Mersey Docks and Harbour Company in 1972. But the Board of 1905 also knew that failing to keep abreast of the times would not only harm Liverpool's vaunted status as "The Second City of Empire" but might even threaten its entire future as a port.[40] The Board was indeed between a rock and a hard place, and the gap through which it might escape was narrowing with the passage of time. Indecision might be as damaging as a wrong decision. As we shall see, the design and early stages of construction of Gladstone Dock exhibited a deplorable degree of irresolution, and it was fortunate that the consequences were not worse.

In reality, however, the Board was living not only on borrowed money but on borrowed time. This might seem an extraordinary statement in light of the way in which it had been held up by the Royal Commission on the Port of London as a model, but in fact all the Commission really showed was that London's former system was even worse. The fundamental problem was that shipowners were either unwilling or unable to pay a price for port services which reflected the true cost of their provision, a figure which rose rapidly with the growth in ship sizes outlined above.[41] As Jackson has shown, this affected all the major British ports severally and individually. In Liverpool its long-term nature and effects were particularly marked.

It all began with the New North Works under the 1873 Dock Act, which cost a little over £2,500,000, all of it, of course, borrowed. The rate of interest on new bonds at the time varied between 3.5 and 4.25 percent. The total debt of the Board before the works commenced was just under £17,000,000, so the additional debt was approximately fifteen percent of the total. The works were completed in 1885, but instead of rising, total revenue fell because the Board had

---

[40]By most logical criteria that honour should in fact be accorded to Glasgow.

[41]This issue is considered in more detail in chapter five.

been forced to cut its rates to meet the threat of the Manchester Ship Canal. Tonnage did rise, but certainly not by fifteen percent and, crucially, the percentage of dues revenue spent on interest payments rose sharply between 1883 and 1886, reaching seventy-eight percent in the latter year.

That is the measure of the situation of the Board when it was attacked for generating an operating surplus which it applied to a sinking fund and to the "Unappropriated Receipts Account." The port's customers seem to have thought this latter a particularly iniquitous impost which should have been applied to the reduction of dues. As will repeatedly emerge below, the Unappropriated Receipts were scrupulously applied to *bona fide* port improvements, and by avoiding the need to go to Parliament for borrowing powers for minor capital works, a considerable amount of money was saved. Much of the cost of dredging the bar and the outer channels, for example, for which the customers had been baying for years, was met from the Unappropriated Receipts. An extremely effective short-term improvement scheme at Queens was funded from the same source. While the Board may sometimes have acted unwisely or indecisively, including its part in the design of the New North Works, no example has yet come to light of irresponsible behaviour regarding the long-term future of the Estate. Its customers seem to have expected it to smash its piggy bank and rush out to buy sweeties for everyone, and then to borrow some money to buy more. In 1895 and again in 1896, persistent pressure brought about decreases in the rates of dues.

The New South Works under the 1873 Act cost about £1,500,000 and proved rather better value than the New North Works, largely because they were better designed for their purpose and therefore did not become obsolete nearly as quickly. That was the truly dire problem at the north end: after a string of minor corrections and improvements came the £1,600,000 Sandon-Huskisson scheme; when Gladstone first was contemplated, the engineer's men were a little over half way through the £3,500,000 works under the 1898 Act. Under these circumstances there was no possibility of the Board significantly reducing its debt, meaning that long-term solvency could only be maintained if there were continuous growth in dues income and no significant long-term rise in interest rates. In 1905 that still seemed possible, but there were certainly those on the Board who were well aware of the dangers. The situation was a direct precursor of the one which finally destroyed the Board in 1972: by the mid-1960s the accumulated debt was some £65,000,000 and, just as in 1905, the Board had to face the choice between losing traffic through offering second-rate facilities and gambling on building a large new dock. The second time around they failed, but before we shower too much praise on their successors who completed Gladstone, we must remember that the cost of Gladstone became an aggravating factor: it was a considerable addition to the historic debt problem faced by the Board in the 1960s.

Such was the stage on which the story of the construction of Gladstone was to be played out. The backdrop was extremely sombre.

Chapter 2
Gladstone Dock

On New Year's Day 1902 the Docks and Quays Committee of the MD&HB addressed a request to the Works Committee that it consider preliminary steps for utilising the land to the north of Hornby Dock to provide accommodation for vessels of up to 1000 feet in length.[1] Reference back to chapter one will reveal that this was a very large size to be discussing before *Lusitania* and *Mauretania* were even contemplated, much less designed. The largest ship in the world at the time was the White Star Line's *Celtic* (1901), of 20,904 gross register tons (grt) and 681 feet in length. Two days later, the Works Committee instructed A.G. Lyster, the Engineer-in-Chief, to show how this land could best be utilised for "vessels of the largest conceivable dimensions." The change in wording is interesting, for when the Board first approved the drawings for the entrance lock it was only 870-feet long: its lengthening to 1070 feet came much later and after many changes of plan, of which more below. This could be taken to indicate that from the start the idea of a ship of 1000 feet was to some members larger than "the largest conceivable dimensions."[2] That must remain in the realms of speculation, however, because the Works Committee Support Papers which would confirm or deny it do not survive for that year. It was, however, normal for committee clerks to recite back inter-committee requests *verbatim* when they came to write up the minutes. At the Works Committee meeting of 7 February Lyster came back with some preliminary sketches which made it clear that for a facility of the size envisaged the land the Board already owned was insufficient, and the Solicitor was instructed to "take preliminary steps" to acquire some fifty-three acres that Lyster had identified as necessary.

---

[1]Minutes referred to but not specifically cited in the rest of this chapter may be conveniently found in National Museums Liverpool(NML), Maritime Archives and Library (MAL), Mersey Docks and Harbour Board Records (MD&HB), Worked-up Papers (WUP) 118/1, where they are sorted by subject and transcribed chronologically.

[2]It did need a certain amount of vision to imagine the continuing acceleration of growth: *Celtic* was well under double the tonnage of *Campania* of 1893, while *Aquitania* (1913) was far more than double the tonnage of *Celtic*.

The need for Gladstone was obvious to the members of the Board, but they kept their intentions to themselves for two very sound reasons.[3] Since the announcement of the scheme would cause a rapid rise in the value of land in the vicinity, as such announcements always did, secrecy must be maintained as far as possible until the Board acquired all the land it needed, and minor purchases continued into 1905 when, as mentioned above, some small-scale preliminary work and some large-scale site clearance had already begun. In 1903, for example, the Finance Committee Minutes are quite specific in recommending deferral of initiating a Bill "pending the result of certain negotiations the Solicitor is conducting for the purchase of land and premises at Seaforth."[4] It is perfectly possible that this was the unspoken reason why the Special Works Committee meeting of 5 October 1903 had deferred the matter as well, ostensibly for further plans to be prepared. It was recognised practice at this date to "censor" the reading of the minutes of the previous meeting by marking items "not to be read in the presence of the reporters;" such items were only read later, after the reporters had left. Minutes relating even peripherally to property purchases are almost invariably so marked.

This was quite understandable: the members (or at least those who knew about it) would remember with special bitterness the way they had lost a significant sum of money to a syndicate of land speculators which included the brother of Edward Gittins, then the Secretary to the Board.[5] The *modus operandi* was simple: information leaked by Gittins was used by his brother to identify land on which options could be purchased before anyone else knew about its potential value. But even discounting such activities, prices invariably rose. In 1904, for example, the Board was buying land which if used for housing would have been worth about eleven shillings per square yard at prices which rapidly rose to 37s 6d, and that was well before the size and scope of the scheme became public knowledge. The rise was simply the result of the realisation that the Board was up to *something* and a shrewd guess as to what that something might be.[6] One did

---

[3]Francis Hyde's statement that "by the end of 1903 the chairman was able to give details" is misleading in two respects: the people to whom he was *willing* to give details were very few in number, and the "details" at his disposal were, as we shall see, both scanty and subject to several subsequent changes. F.E. Hyde, *Liverpool and the Mersey. An Economic History of a Port, 1700-1970* (Newton Abbot, 1971), 125.

[4]NML, MAL, MD&HB, Finance Committee, minutes, 13 November 1903.

[5]For details of this near-criminal breach of trust, see my forthcoming paper "Corruption in the Port of Liverpool." The speculators' most successful coup yielded them some £47,235 profit, entirely at the expense of the Board, in an afternoon.

[6]NML, MAL, MD&HB, Discussions at Board, 2 June 1904.

not have to be terribly prescient to do this: throughout 1904 and into 1905 Sir John Jackson Ltd.'s men were busily removing the old spoil bank which lay to the north of Hornby Dock, involving some 1,367,000 cubic yards. Clearly something big was about to happen.

The second reason for keeping quiet was that if the scheme was announced a significant length of time before the Board could implement it, the perilous state of its finances might become known, whereupon the well-known financial principle "and from him that hath not, shall be taken away even what little he hath" would come into play, severely damaging the prospects of raising the money until there had been a major upturn in the port's fortunes - and who knew when or even whether that might happen, without the new dock? As we shall see, just such a delay occurred, and its consequences interacted with a period of slack trade with very unhappy results.

There now followed a period which must have been infuriating for Lyster and his department. On 12 February 1904 the Works Committee discussed the general layout of the new works and instructed Lyster to produce a north-south-oriented scheme similar to that at Toxteth and Harrington docks but with an entrance lock like the new one to be built at Brunswick, which was at a very narrow angle to the river to allow shipping using it virtually to stem the tide as it entered or left the river.[7] He duly produced plans complying with varying briefs on 27 July, 2 September and 17 March 1905, and 12 May, 26 May and 16 June 1906. The cost of this exercise in indecision must have been enormous, though in the defence of the Committee it must be conceded that it rightly thought that getting it wrong would be a truly epoch-making disaster in the history of the port.[8] And it knew how things could go wrong, both in the drawing up of the brief and at the design stage. Lyster had been responsible for a few errors, including building the quays at Queens Dock too high for the class of ship that used them: his father and predecessor had perpetrated some designs of real folly, some of them in response to a brief so badly-drawn that it could elicit no better.[9]

The Works Committee at last felt happy enough with the proposal to recommend on 23 June that it go to the Parliamentary Committee. It had so evolved that it was quite unrecognisable from the original scheme, in that it was

---

[7]This arrangement has (it is still in use) the further advantage that a certain amount of windshielding, to protect ships from onshore winds, could be provided.

[8]Once again, Hyde, *Liverpool and the Mersey*, is misleading, probably in the interests of brevity, when he stated that "the board acted with speed and efficiency." It demonstrably acted with neither.

[9]For details of some of Lyster Sr.'s mistakes, see A. Jarvis, "G.F. Lyster and the Role of the Dock Engineer 1861-97," *Mariner's Mirror,* LXXVIII (1992), 177-199.

now a "vestibule and branch" dock, shaped like Alexandra, not a north/south rectangular one, and with a half-tide entrance rather than a lock like the one at Brunswick. But if Lyster thought he had seen the back of it for a month or two he was wrong: on 14 July, in the Works Committee "[a] further discussion having taken place on the approved plans...the Engineer was instructed to produce a new plan" showing what would be needed to handle ships up to 1000 feet in length. This had, of course, been a basic parameter in the very first mention of the scheme by the Docks and Quays Committee.

On 20 July 1905 the Board confirmed the recommendation of the Parliamentary Committee that the Bill go to Parliament in the next session. The details of the scheme (as it then stood) were released to the press on 7 July, where they received extensive coverage and favourable comment. Once the intention of building the dock and some details of its nature became known, the customary process of seeking support from the customers who would or might eventually give evidence in favour of the Bill began, as did the task of placating or avoiding as many of the potential objectors as possible. From the background given above, one might expect that there would be a good deal of objection from those trades which used medium-sized vessels and would resent the expenditure of the money they paid in dues on facilities built chiefly to suit a handful of exceptional ships owned by just two companies. Surprisingly, this did not happen, and when Robert Gladstone came to give evidence before the Select Committee on the Bill of 1906 he was able to provide a list of supporters which was impressive in its catholicity.

The support of the Liverpool Steam Ship Owners' Association comes as no surprise, but the Liverpool [sailing] Shipowners' Association might have been expected to see little benefit in a huge new dock for steamers. The sail tonnage using the port was declining fast, but in 1906, 172 foreign-going sailing vessels with an average register tonnage of just over 1000 entered the port, together with 3188 coastwise, with an average tonnage of just over eighty. The total tonnage entered was 433,294, a not insignificant amount of traffic if we recall that total entries for sail and steam together for the same year in the supposed "competitor" port of Preston stood at 255,603.[10] The Liverpool sailing ships might now only gain the crumbs that fell from the steamers' table, but if the steamships moved to better berths, those they vacated would be better than the sailing ships' existing accommodation, and the Association had more sense than to expect that new docks would be custom-built to suit sailing vessels.

Given the long-standing conflict between the interests of the shipowner and the merchant, both within and without the meetings of the MD&HB, the bill

[10]D.J. Starkey (ed.), *Shipping Movements in the Ports of the United Kingdom 1871-1913: A Statistical Profile* (Exeter, 1999), 154-55 and 200-201.

produced a surprising measure of harmony.[11] The Incorporated Chamber of Commerce of Liverpool, which had made some ferocious attacks on the MD&HB in the past, was supportive, as were a number of trade associations. Again, we might not be too surprised to find the Cotton Brokers' Association, whose members' goods typically came in some of the larger cargo steamers using the port, in support, but the same could not be said of the Fruit Brokers' or the Provision Trades' associations.[12] According to Hyde, the timber trade opposed the Bill - which the minutes of evidence show they did not. Like the grain trade, they merely failed to support it. It is true that both trades spoke against the proposal within the confines of the Board and, no doubt, in private meetings. They recognised, however, that the extension of the port was essential and, like the sailing shipowners, they could benefit from good deep-water berths vacated, for example, by Cunard in Canada and Huskisson docks. This is possibly the first time that the Board's customers were willing to sink their differences in the larger common interest. It is in very marked contrast with the behaviour of the representatives of the Irish Trade, who had opposed the 1898 Bill out of sheer jealousy and frustration at the Board's decision to improve passenger liner facilities at the floating stage while doing nothing new to help the coastwise passenger trade.

The evidence given on behalf of the Board itself was eminently predictable and met with little dispute. John Hughes, Chairman of the Works Committee, produced tabulated information showing the rapid growth of trade and of average ship sizes, and explained to the Select Committee why there was a pressing need for more appropriated berths. He described as "a great delusion" the often-heard suggestion that appropriated berths were an inefficient use of space.[13] A particularly interesting piece of evidence is a set of fold-out graphs tipped into the bound version of the bill. It plots the recent increase in size of passenger liners and extrapolates it to 1909. It gets *Lusitania* right, because its approximate dimensions were already known, but, surprisingly, if extrapolated further it gets very close to forecasting the size of *Aquitania* as well.

---

[11]For the roots of conflict, see G. J. Milne, "Port Politics: Interest, Faction and Port Management in Mid-Victorian Liverpool," in L.R. Fischer and A. Jarvis (eds.), *Harbours and Havens: Essays in Honour of Gordon Jackson* (St. John's, 1999).

[12]Though the average tonnage of steamers using southern US ports was only 6124, as against 12,213 in the New York and Boston trades, this was still larger than ships in any other trade except the Canadian. NML, MAL, MD&HB, P1/97.

[13]This is a view which the "efficiency" figures do support: docks which were mostly occupied by appropriated berths tended to have larger throughput - but they also tended to be the more modern docks, so they should have been more efficient. Perhaps Mr. Hughes' position as senior partner in the Harrison Line affected his view.

The scheme as described by Lyster consisted of two 130-foot-wide entrances with a depth of thirty feet below Old Dock Sill (ODS), which would give about forty-five feet at mean high water, leading into a vestibule or half-tide dock which was also to have a ninety-foot passage to Hornby Dock. Two parallel branch docks, more or less as built and still in service, were to have 120-foot entrances and a depth of thirty feet below ODS. The total estimated cost of the scheme was £4,027,500.

Virtually all the opposition to the Bill came from local authorities. Liverpool City Council wanted protection for its water supplies to the Dock Estate – which was no trifling matter, as the Corporation's published accounts for 1907 show charges for supplying water to ships as raising over one-third as much as the entire domestic water rate. Bootle wanted protection for its sewerage system, one outfall of which crossed the proposed site. Wallasey Urban District Council objected to a proposed power of the Water Bailiff to break up "shipping casualties" on the foreshore. The Lancashire and Yorkshire Railway wanted to preserve its access to a plot of land with foreshore rights which it owned adjacent to the site. Crosby Urban District Council argued with perhaps some justice that dockland activities would lead to depreciation of the residential property nearby – but it was not actually all *that* nearby. Waterloo-with-Seaforth Urban District Council tried neatly to have its cake and eat it by arguing both that nearby residential property would lose value and, at the same time, that land belonging to the Council which was suitable for dockbuilding purposes would also lose value if Gladstone were built. This is a truly classic example of a public body contemplating squandering its funds in a perfectly pointless attempt at competition with a major port. The figures from Preston, mentioned above, might have given them pause for thought, but probably would not have done, because what they really wanted was that MD&HB should be compelled to buy their land at a high price.

Two different surveyors and valuers, Mr. Hartley and Mr. Wainwright, gave evidence that property values in Waterloo had been sliding for at least twenty years and that the effect of dockbuilding was to raise, not lower, the value of land, but that is not the important part. What matters is that on engineering and commercial grounds the bill was, in effect, unopposed.

In happier financial circumstances, the passage of the Act would have been the cue for Lyster to hire several hundred "extra labourers" and send them off to start digging.[14] Instead, there was almost complete inactivity: the Board's published accounts show expenditure under the 1906 Act at £2237 in 1906 and £3864 in 1907. This, again in contrast with the received view, allowed time for

---

[14]The term "extra labourer" has a specific meaning in the context of the Dock Engineer's Department: they were technically casual workers, but in practice the same men (unless dismissed for misconduct) would be retained until the work or the money ran out.

more self-doubt and vacillation. On 6 November 1906, for example, Lyster was instructed to produce a new plan, with estimated cost, "for the longest lock which could usefully be provided from the river to the proposed docks." This was, of course, a reversal of the decision to have a lock resembling that at Brunswick. On 4 April 1907, the request was for "[t]he lock leading from the river to the half-tide dock to be, if possible, 870 feet in length." This envisages an arrangement similar to that at Sandon, where locks known to be of insufficient length for exceptional vessels were so arranged as to allow half-tide working for those vessels, while others used them as normal locks. Even by the sometimes low standard of policy-making exhibited by the Board, this was indecision of memorable proportions.

It was not as if there was a lack of professional literature to guide them. The Institution of Civil Engineers, The Institution of Naval Architects and the Permanent International Association of Navigation Conferences (PIANC) published papers recording recent growth in ship sizes and attempting to foresee the nature and extent of future growth. PIANC's 1905 meeting in Milan and the ICE Engineering Conference of 1907 both addressed this as a major issue.[15] Lyster, or at least members of his staff, would of course be familiar with this literature. It will be noted that, in the sequence of requests for plans above, there is no mention of seeking his advice as to size: in every case the size of vessel to be provided for was laid down by the Board, whose members were quite possibly not familiar with the literature. The issues are well covered in the Discussions at Board sequence of *verbatim* minutes, where no mention is made of any such reading, and the tone of discussion clearly indicates that members were simply voicing their own opinions, based on experience followed by fairly well-informed guesswork. This might be thought a manifestation of the "we know best" attitude which was very noticeable in the discussion of any issue which involved comparing Liverpool's way of doing things with the practices of other ports.[16] But while employing the wrong method increases the risk of getting the wrong answer, it does not preclude the possibility of getting the right one. In the event, even if it was only by a happy corporate guess, the right long-term decision was made, but only after further agonising and after a fairly unhappy short term.

---

[15]See, for example, Institution of Civil Engineers, Engineering Conference, 1907, *Report of Discussions* (London, 1907).

[16]See, for example, NML, MAL, MD&HB, Discussions at Board, 15 January 1891. This was a discussion of the report of a Special Committee which had been investigating the Board's general methods of doing business. There are some good ideas here, forming a useful basis for reform, but also some statements of staggering smugness. A retired principal officer of the Mersey Docks and Harbour Company, who joined in the last days of the Board, informs me that the idea that one could learn from others was still largely alien in his early years in the job.

In a memorandum of a discussion which Mr. Hughes, the Chairman of the Works Committee, had with the Finance Committee on 25 February 1908 we find deeper doubts even than those expressed previously. The Board had been shocked by the recent decision of White Star to move its express passenger services to Southampton and feared that Cunard might follow.[17] During this meeting it was reported that

> The Chairman of the Board strongly urged the unremunerative nature of the outlay if it were merely to cater for passenger boats [sic] of the type of *Lusitania* with their "fictitious" net tonnages...This expenditure would be such a burden nothing could be done with modernising the central docks or the landing stage.[18]

In this as in many other matters, Mr. Gladstone's judgement was sound, at least in the short to medium term, but it is alarming that such uncertainties prevailed at so late a stage. Stranger things were still to come. After the Board meeting of 23 April an informal discussion was held on the subject of the "north end extension," where matters headed towards becoming so confidential that members queried whether it should be recorded, and the minutes end abruptly with the words "The shorthand writer withdrew."[19]

The *Journal of Commerce* for 1 May 1908 reported Hughes' proposal that work should begin, and on 8 May the start was authorised, though Lyster was firmly enjoined not to spend more than £250,000 within the calendar year. On 12 May his men began fencing the site and demolishing the waste destructor, which was to be replaced with a new one in Lightbody Street.[20] By August the financial situation had deteriorated again, and the Finance Committee instructed

---

[17]This was more a blow to the ego than the wallet: NML, MAL, MD&HB, Steamship Owners Returns, 1908, show White Star paying more, not less, in dues to the Board than in 1906. (Though of course these figures were not yet available when Mr. Hughes spoke in February.) It was actually the "bread-and-butter" ships of both White Star and Cunard which generated the revenue, and the eventual removal of the Cunard express services to Southampton in 1919 had equally little effect.

[18]This refers to the fact that *Lusitania*'s net register tonnage was only twenty-eight percent of its grt, meaning that, in the Board's eyes, it paid less than one-third of the dues it should. NML, MAL, MD&HB, WUP 118/1.

[19]*Ibid.*, Discussions at Board, 23 April 1908.

[20]There is a detailed account of this installation in *Engineering*, 17 November 1911, 666-668.

that Lyster's weekly expenditure, which was running at a modest average of about £1000 per week, be reduced to under £500. On 21 August he was given authority to spend a little more in order to keep "about 160 men" engaged in preparatory works. When in October he asked for some more, he was refused. The axe fell on 11 December 1908, when he was required to produce a report "stating how he proposes to dispose of the men at Gladstone Dock Works *which it has been decided to discontinue*" [my italics], and a week later he was instructed to give them notice. The total spent during the year ending July 1908 was a trifling £7214, followed in the 1909 *Annual Accounts* by just £3932.[21]

The last step in this particular strand of the story came on 30 March 1909, when it was reported to the Finance Committee that Lyster had privately compiled and printed a pamphlet urging continuation. This, according to prevailing protocol, was highly impertinent. It roused the members to righteous anger, and they instructed Lyster that the works were "not to be proceeded with" for at least twelve months. Within those "twelve months" - which in fact proved to be many more than twelve - much would change. Expenditure in the year to July 1910 was the princely sum of £138, apparently the wages of keeping a watchman on site to mind the plant that Lyster had left in anticipation of restarting the works.

We need not look far for the reasons. The first and most obvious was that a fairly good year in 1907 produced dues income for the Board of £1,358,884, but that was followed by three bad ones. Freight rates were recovering only slowly from their plunge between 1900 and 1902, and the general level of trade was not encouraging. On 27 March 1907, £323,000 of the Board's bonds fell due and only £87,000 were converted or renewed.[22] At the Board meeting of 18 April it was estimated that "we shall have a deficiency on 1st July of probably £500,000." Interest charges that year amounted to 66.5 percent of dues income, and in 1908 this rose to 68.7 percent, 69.9 percent in 1909 and seventy-one percent (the highest level since 1892) in 1910. In that year the Board's total capital expenditure amounted to £46,720 as against unexpended - but unaffordable - borrowing powers of £5,224,366.[23] No wonder that Lyster was

[21]There was an element of screaming before pricking their thumb here, but in fact the members had guessed the trend correctly.

[22]NML, MAL, MD&HB, Discussions at Board, 28 March 1907.

[23]One of the measures of the financial soundness of a port which was bandied around at that time was the amount of unexpended borrowing powers held, and another was the difference between actual dues charged and the maximum allowable under statutory powers. Both were highly misleading because they neglected to take account of the fact that a port can only charge what the market will bear and can only borrow what it can pay for.

not allowed to resume work until a marked upturn in trade took hold in 1911. The situation was further improved by the fact that in December the Board rightly chose to increase its dues in a rising market, resulting in an increase of dues income from £1,290,480 in 1910 to a much healthier £1,469,409 in 1912. Most important, the interest expenditure had fallen sharply to 62.9 percent of dues income.

There was also scope for a certain sense of relief in that it finally seemed that a solution to the problem of the bar was in sight. The Board owned the two largest dredgers in the world, and the completion of the Taylor's Bank revetment promised to prevent recurrent shoaling of the channel, allowing the dredgers to get the whole system under control. In fact ,such hopes proved ill-founded, but a "feel good factor" existed, if only briefly. The Board had also managed to shed the ill-fated Dinorben Quarry in Anglesey, which it had leased and used to provide stone for the revetment.[24] The Brocklebank Branch was finished, as were the last few loose ends from the works under the 1898 Act at the south end. It was a time when at least some of the members could begin to feel a bit optimistic again.

The observant reader should have noticed a significant omission in the narrative: there is no mention of any provision for a graving dock. It is true that North Atlantic liners used state-of-the-art antifouling paints and that their high speeds, short periods in port and their confinement to cool waters made them much less prone to fouling than most vessels.[25] It did not, however, make them exempt. No specific information on docking frequency has come to light, but the Royal Navy's guideline was that a large vessel in the North Atlantic lost .25 percent of its overall efficiency for every day since it emerged from the dock. Fouling was a serious, if under-recorded, matter.[26] In addition, it must be remembered that express liners often explored the boundaries of technology and

---

[24]For the story of dredging the bar see A. Jarvis, *The Liverpool Dock Engineers* (Stroud, 1996), 83-89; chapter six, below; and J.A. Cashin, "Engineering Works for the Improvement of the Estuary of the Mersey," *Journal of the Institution of Civil Engineers*, XXXII, No. 7 (1948-1949), 296-367. On Dinorben, see A. Jarvis, "The Mersey Docks & Harbour Board in Anglesey: Dinorben Quarry," *Maritime Wales,* XIX (1997-1998), 93-99.

[25]Some, however, were already engaged in pleasure cruising in warmer waters during the winter.

[26]The latest information at the time was in A.C.A. Holzapfel, "Ships' Composition," *Transactions of the Institution of Naval Architects*, XLVI (1904), 252-272. For a summary of views, see A. Jarvis, "Protection and Decoration: A Tentative Investigation into Painting Ships before the Great War," *The Great Circle*, XXII (2000), 24-37.

in so doing could suffer teething problems underwater. It was unusual, but by no means unheard of, for propellers to work loose, or even to fall off completely *en route*, and rudders and rudder frames were not indestructible, either.[27] In short, running a prestige express liner into a port which could not accommodate it in graving dock was a risk the average shipowner would be unlikely to take.

The largest graving docks in Liverpool at the time the original Gladstone project was being developed were those at Queens, Canada and Brocklebank docks. None was adequate for ships of the size envisaged by the Docks and Quays Committee in 1902: Queens was only 634-feet long, and Brocklebank was still on the short side at 804 feet. The largest graving dock was Canada at 925 feet in length which had a ninety-two-foot-wide entrance. Its sill, at twelve feet, eleven inches below ODS gave only about twenty-eight feet at mean high water, though in fairness that section of the docks was served by impressive impounding pumps capable of maintaining the water at the level of high water spring tide (HWST), another three feet. It was a very good dock, suitable for the largest ships in service when the Gladstone Enabling Act was passed, but not adequate for the future monsters which were already being envisaged.

When Joshua Ramsbotham read a paper to the Liverpool Engineering Society on the construction of Brocklebank Graving Dock he was given a slightly hard time in the discussion. But of the various criticisms made of the scheme, not one suggested that it was of insufficient size.[28] While that might be expected in a gathering comprised entirely of civil engineers or dock specialists, this was not such a group: one of those involved in the discussion was J. Reney Smith of the prominent local ship-repairing firm of H. and C. Grayson and Co., a major user of the graving docks and a member of the Institution of Naval Architects. Among members not present at the meeting was a noticeable minority of naval architects, some employed by shipping lines, who could have sent in written comments. But the only written intervention came from a Belfast harbour engineer. In short, Ramsbotham's paper does not reveal any uncommon degree of myopia in the matter of the effect of ship size on graving docks.

There were, on the other hand, those who recognised the implications of published work such as Lord Brassey's paper on "Merchant Cruisers and

---

[27]For example, *Mauretania* had two lengthy stays in Canada Graving Dock in 1911 as a result of an anchor cable failure that allowing it to drift onto Pluckington Bank. Some 700 tons of new plates were necessary. For this and similar misfortunes, see N. McCart, *Atlantic Liners of the Cunard Line. From 1884 to the Present Day* (Wellingborough, 1990), 63-64.

[28]J. Ramsbotham, "Graving Dock Construction, Dealing with the New Brocklebank Graving Dock," *Transactions of the Liverpool Engineering Society*, XXVIII (1907), 45-70 and 101-116.

Steamship Subsidies."[29] In essence it was suggested that purpose-built naval scout vessels were a waste of money because they could never be as fast as passenger liners since they could not be big enough, so fast liners should be subsidised in order that they could be requisitioned in the event of war. It is perhaps worth digressing to consider this particularly nonsensical argument in favour of spending public money subsidising ships "of uneconomic speed," as *Lusitania* and *Mauretania* were called before they were built.

A number of purpose-built scout vessels went into commission in the years 1904-1906. They resembled light cruisers in being typically about 3000 tons with a top speed of twenty-five knots. They had a modest amount of armour and were normally armed with ten twelve-pounder guns and a few torpedoes. They were later given four-inch guns and re-designated as light cruisers. For the sake of an extra knot, a 30,000-ton ship, at full speed burning up to 1000 tons of coal per day, designed to coal in port once a week, was supposed to be a more a suitable fleet scout.[30] Although they might be fast, their lofty superstructures made them recognisable before they could identify enemy units - and easy to hit with long-range gunfire. Nor were they necessarily fast enough to run away - new destroyers of the day were typically capable of thirty-three knots, even if only in short bursts and fairly calm water. The liners were specialist tools, beautifully and carefully designed to regain the Blue Riband and maintain British seafaring prestige: as such, they were fairly useless for anything except the carriage of troops.

Furthermore, in the event of war such vessels might well be in even more demand for maintaining normal communications, transporting troops and *materiel* or acting as hospital ships. Despite a slight economic downturn and soaring naval estimates, the government needed to build battleships to maintain the two-power standard, so subsidy of large merchant vessels seemed to some to provide an answer to the fleet scout problem. Once that excuse (among others) had been provided, the arrival of monster ships was assured. It was also assured that government would not subsidise the new facilities they would require.

This being so, it seems quite extraordinary that nobody, not even the Liverpool Steam Ship Owners' Association, objected to the 1906 Bill because of the absence of a giant graving dock. No less strange is the fact that not one of the re-designs mentioned above involved Lyster providing a graving dock. The only tentative conclusion at which we may arrive is that, whatever may have been said

---

[29]Lord Brassey, "Merchant Cruisers and Steamship Subsidies," *Transactions of the Liverpool Engineering Society*, XXVIII (1907), 30-38.

[30]By way of comparison, a dreadnought steaming at its maximum speed of twenty or twenty-one knots did so on about one-quarter of *Lusitania's* horsepower and about half its coal consumption.

about 1000-foot ships, it was anticipated that Canada Graving Dock would prove sufficient for some considerable time. That implies that the leap from *Cedric* to *Lusitania* had been foreseen, but not that from *Lusitania* to *Titanic* or *Aquitania*. Since the White Star express liners had moved to Southampton, we may connect the need for a graving dock larger than that at Canada with just one vessel - *Aquitania*.[31]

The scheme for a vast new graving dock which would also serve as a wet dock for a single enormous ship first came before the Board in June 1910. It was one of two solutions Lyster put forward to the problem that one particular shipping company (Cunard's name was coyly withheld at this stage) was "about to build a very large vessel somewhat in the style of the *Mauretania*."[32] His first option was to enlarge and improve the berths and graving dock at Canada, which was really no option at all. Repeated attempts by Lyster, and his father before him, had failed to provide sufficient depth at Canada Entrance, and to improve on what was there in 1910 it would have been necessary, among other things, to demolish the huge concrete apron which Lyster Sr. had provided as part of the silt sluicing system. The estimated cost was £1,000,000, and anyone who knew the track record of works at Canada Entrance would have expected considerable overspending. It is conceivable that the entrance at Sandon could perhaps have been improved again, but not quickly or cheaply: the 1891 works had taken ten years, caused a significant drop in the efficiency of the whole North system while in progress and cost the large sum of £20,000 per water acre - almost the price of building from scratch. The access from there to Canada Graving Dock was not easy and would require further passage enlargements, so "the 'improvement option" wisely was not seriously considered.

Lyster's preferred solution was to make some relatively minor - and one very major - changes to the Gladstone dock plan. The half-tide dock would remain, even if it ended up as a fully impounded dock, to serve as a "vestibule" from which ships could swing round to the main lengths of quay on the two branch docks. To the north of No. 2 Branch was to be the great dual-purpose graving dock. It was relatively cheap, at under £500,000, and it could be built on its own, without having to complete the rest of the scheme until later, meaning that it could be in service within three years. It had significant disadvantages,

---

[31]In 1913 there was a flurry of speculation in the press that the Hamburg-Amerika liner *Imperator* might need to use it. It was hardly likely at that time to become a regular customer, though it did in fact use the dock in its new persona as *Berengaria* while still under requisition.

[32]Why they were so coy is unclear: the Works Committee had discussed a report on further graving dock accommodation with representatives of Cunard on 13 May. NML, MAL, MD&HB, Discussions at Board, 30 June 1910.

though. Some £180,000 of the estimated cost was for its temporary river entrance, which would, of course, no longer be required when the main scheme was implemented. Moreover, the entrance and its approach channel would almost certainly require frequent and costly dredging, and the need to cope with a differential head to either side of the entrance gates encouraged the employment of a rolling caisson gate.[33]

Some members of the Board, encouraged by Hughes' statement that dock revenues were improving and that it should soon be possible to borrow money again, became enthusiastic about pressing on with the main scheme at the same time. Mr Cunningham even went so far as to suggest that the Board should approach the City Council with a suggestion that it should emulate Manchester City Council's support for the Ship Canal and buy large quantities of debentures. The circumstances were not really comparable and this, like other similar suggestions, received the hostility it deserved.[34] Suggestions were also made that if the Board were to spend all this money simply for one company, there had better be some sort of agreement to bind Cunard to the port.

The Board, after its long period of enforced inaction, now moved with considerable speed. On 30 June it instructed Lyster to proceed with a dual-purpose dock "about 1020 ft. long" with a 120-foot-wide entrance and a sill laid at twenty-nine feet below ODS, corresponding to about a forty-four-foot depth at mean high water. The "fairway" to the entrance was to be dredged to twenty-nine feet below ODS. He was to provide a single-story shed, 900 by 100 feet, on the North Quay and four thirty-hundredweight movable cranes and was to do it "with all practicable despatch." The frighteners were well and truly on, though perhaps not sufficiently to overcome the Board's sense of its own importance. At the Works Committee meeting of 7 October it was minuted that the General Manager had drawn attention to the fact that the Clyde Navigation Trustees had taken the decision to build a 1020 by 100 foot graving dock and that the local press was publishing "eulogistic" reports of how it would be the biggest in the world. In fact, it would not have been as big as Gladstone anyway, but Lyster was sent off to find out how much it would cost to extend Gladstone "by either ten or twenty feet." This uncertainty as to size was still matched by a lack of final decision as to shape: as late as 30 December the Works Committee agreed to send drawings of a modified section of the dock to Cunard "for their official approval," which

---

[33]Saving on that maintenance cost later became one of the arguments for proceeding with the main scheme. For problems with the caisson gate, see below.

[34]If Mr. Cunningham (who was scarcely a newcomer, having joined the Board in 1895) really thought the members would accept corporation representation on the Board, he had not been paying proper attention in meetings. There was a definite antipathy, albeit usually fairly well concealed, between the two bodies.

was received in a letter of 7 January 1911. In the event it was built 1050-feet long, while the Glasgow dock was never built.

Nor had all the financial doubts been banished, as discussion at the Board meeting of 7 July 1910 shows:

> After the conclusion of the ordinary business a long discussion took place with regard to the Board's general policy in connection with the construction of New Works at large cost and the probable inadequate return thereon and also as to the Board's methods of providing for such expenditure, and the Board's financial requirements generally.

This is a curious minute in that it does not define "adequate return." Indeed, one might argue that the continual increase in the Board's debt over its fifty-year life indicated that revenues had always fallen short of the combined costs of interest and obsolescence and hence of an "adequate return."

That meeting had, however, taken the formal decision to proceed, and Lyster now had to get work moving as rapidly as possible. It was decided that in the interests of flexibility the work should all be done "in house," and excavation began on 19 September 1910. This was an interesting decision, since Lyster had been directed to submit a departmental tender in competition with private contractors to build Brocklebank branch and had taken some severe criticism over the in-house quarrying venture at Dinorben. The only thing that seems to have changed is the degree of urgency to have the new dock ready before *Aquitania,* and in those circumstances an in-house job seemed to offer a better chance of completion by the due date. Whatever the reasons, it turned out to be a correct decision.

Spoil raised from the site of the dock was used to form an embankment to enable working "in the dry." The urgency of the job was again reflected in the employment of over 700 men working a two-shift system, and the considerable task of forming the embankment was completed in five months. By July 1911 Lyster was able to report that the excavation of the trenches in which the dock's walls were to be built was nearing completion, that the "curved head of the dock" was completed and that both the north and south walls were advancing westwards.[35] It is perhaps worth emphasising the size of the task: the trenches were fifty-feet-deep from shore level, the walls were sixty-feet high and their maximum thickness was nineteen feet six inches. This intense activity did not come cheap: in the nine months' work included in the *Annual Accounts* for the year ended 1 July 1911, Lyster spent £109,288 and in the full year which followed he spent

---

[35]Making the head rounded made it act as an arch in resisting ground loadings, which allowed it to be made significantly thinner and cheaper than if it were square.

another £133,296. This was admittedly fairly small compared with his expenditure when the works under the 1898 Act were in full swing. but it still compares favourably with that of the troubled years of the immediate past.[36] The Board's *total* capital expenditure in the 1910 Annual Accounts was just £46,270.

Lyster's next annual report tells us that "the walls of the dock and entrance are practically completed" and that the excavation of the body of the dock had begun. Lyster was not usually of the school of engineering which boasted about quantities, so he simply mentions some figures: 227,896 cubic yards of excavation and 66,793 cubic yards of concrete and masonry within the year.[37] This was a creditable performance given the loss of "five weeks of the best weather" occasioned by the great transport strike, which came "near to revolution."[38] The effects of the strike were both direct and indirect, in that some of his labour force joined the strike and those that did not were left with no coal available for the considerable amount of steam plant on site.[39] The winter which followed was marked by very unpleasant weather which "caused great delay."

The year 1912 seems to have passed without any major labour problems, but in February 1913, 462 out of the 927 men on site went on strike. The chief issue was the rate of pay of the navvies, but most of the tradesmen's labourers struck as well, and eighty-seven of those not directly involved, chiefly tradesmen and drivers of steam plant, who could not work without labourers, chose to strike

---

[36]For example, £399,317 in 1901 and £443,619 in 1902. NML, MAL, MD&HB, *Annual Accounts*.

[37]He did, however, boast about the prowess of his dredgers, including the statement that the amount dredged from the bar by 1899 would suffice to fill the Thames solid to high water mark from Blackfriars Bridge for a distance of five to six miles. A.G. Lyster, "Works Recently Carried out and in Contemplation at the Port of Liverpool," *VIII International Congress on Navigation* (Paris, 1900), 9.

[38]E.L. Taplin, *Near to Revolution: The Liverpool General Transport Strike of 1911* (Liverpool, 1994).

[39]There was coal physically available, but 800 men were laid off because, while willing to do their own work, they were not willing to handle the coal. The predictable accusations of intimidation were bandied about and may well have been genuine, for it was an extraordinarily bitter strike. It is clear that the Board feared major sabotage: the above was reported at the meeting of 23 August 1911, but it soon moved on to the issue of providing protection for the dock gates. At that same meeting the Chairman presented a copy of a letter he had sent to the Prime Minister explaining the consequences if the Board was unable to "keep the locks in order," and the Board expressed its relief that there was, as it quaintly put it, a "man o' war" in the river. NML, MAL, MD&HB, Discussions at Board, 23 August 1911.

rather than "blackleg" by doing some of their own labouring.[40] Despite these problems, however, it was not Lyster's men who would delay the dock being put fully into commission, for when it was formally opened on 11 July 1913 all their work was complete, while the two main contractors' works were not. When an enquiry as to the availability of the dock for *Imperator* was put to the Docks and Quays Committee on 6 August, it was reported that the pumping engines were not "yet in place."[41] The great rolling caisson gate was in place and operable, but it was still suffering teething troubles.

By 1 July 1913, the dock itself was finished and in water and the caisson gate, 132-feet long, fifty-two-feet six inches high and twenty-five feet "thick" was being built *in situ* and was nearly completed. The pump house was ready to receive the five great diesel-driven pumps, the first of their kind, which would be able to pump out the dock's more than forty million gallons in two and a half hours.[42] On 11 July 1913 the King and Queen entered the new dock on the Board's steamer *Galatea* to perform the ceremonial opening. It had been a remarkable effort, undertaken in the face of economic adversity and the consequent nervousness and indecision of the Board.

It seems clear that one consequence of the speed with which Lyster had moved was to embolden the members, for as early as 16 January 1912 the Finance Committee approved a proposal for the main scheme to go ahead, and on 10 May 1912 he was instructed by the Works Committee to recommence work on it. Preliminary works for the construction of Hornby Lock, to connect the new docks to the rest of the system, were begun immediately, and by July 1913 the walls of the two branch docks were in hand. This was reflected in the expenditure between July 1912 and July 1913 reaching £273,000 which, of course, means that we cannot tell from the published reports whether the graving dock was completed within budget or not. Needless to say, the resumption of work also involved changes in the design. Various bits of "paring down" had reduced the estimated cost from the 1908 figure of £3,204,400 to an encouraging £2,616,200,

---

[40]*Ibid.*, 27 February 1913.

[41]The engines were highly innovative diesel units built and installed by Carels Frères of Ghent. The price of innovation is often, as here, a matter of not getting it quite right the first time.

[42]There is a fairly full description of the dock and its equipment in *Engineering*, 4 July 1913, 4-13, but the definitive accounts are W.H. Jones, "Gladstone Graving Dock, Liverpool," *Transactions of the Liverpool Engineering Society*, XXXVIII (1916-17) 199-210; and W.A. Oglethorpe "Gladstone Graving Dock, Liverpool - Machinery," *ibid.* 211-246. The amount of water in the dock varied according to the level outside and the size of the vessel using it. Pumping dry for test purposes (i.e., starting full of water but with no ship in) used about four tons of diesel fuel.

and on 6 September 1912 the Works Committee approved the excision of the sluicing system intended to clear silt away from the entrance resulting in a further saving of about £64,000. On the debit side, on 12 December it approved the lengthening of the entrance lock from 870 to 1070 feet at an additional estimated cost of £118,000. Presumably this was intended to provide longer hours of access for normal-sized vessels, since the admission of exceptional ships by opening all the gates of the lock at the top of the tide was already possible because the "vestibule" remained capable of use as a half-tide dock, separated from Hornby by a lock.

Be that as it may, it did not take long for the Board's mistake in pandering to Cunard to be revealed. When the graving dock was opened, there were just six ships of over 30,000 tons in the world, to which, at the time the Board was making its decisions, one could notionally add *Titanic* and *Britannic*, which, like *Olympic,* were under construction. Of these, *Imperator* and *Vaterland* were scarcely likely to become valued customers, and the "return" of the White Star vessels was scarcely more likely: certainly it was not put forward in the Board's discussions as an argument in favour of going ahead at Gladstone.[43] That reduced the market to three ships: *Lusitania, Mauretania* and *Aquitania*, of which the first two were already working from the berths at Canada/Huskisson and had successfully used the Canada Graving Dock.

In 1913 the interim Gladstone Dock earned a total revenue of £18,119. Its best year was 1919 when it earned £30,483. During the war years its income never reached £10,000, and at the time of Cunard's departure its life-long income still fell well short of the £180,000 cost of the temporary works undertaken solely to have it ready for *Aquitania*. In its dry-dock role, once the war and its aftermath were over it scarcely ever accommodated a merchant vessel which could not have used Canada, though its depth was occasionally fully exploited by a visiting battleship.[44] Large enough in theory to accommodate either of the Cunard "Queens," it had no centre gates to save on pumping costs when used by lesser craft or to allow it to be used by two vessels, as was the practice in many of the Board's graving docks. Wasting money on a facility like that was exactly what the Board did not need when it had so recently peered into the abyss of insolvency.

---

[43] *Imperator, Vaterland, Olympic* and *Aquitania* all used the dock while under Admiralty requisition, but did not return to peacetime service from Liverpool.

[44] The first battleship to use it was HMS *Erin,* one of two which had been nearing completion in Britain for the Turkish navy at the time war broke out, when they were seized by the Royal Navy. A battleship's armour makes its "light" draught a much larger proportion of its "fully laden" draught than is the case with merchant vessels, causing it, size for size, to need a deeper dock.

Summarised like that, the story sounds altogether more straightforward than we have come to expect of the Board's business, and so it is. To begin with, the caisson gate was troublesome, and it was not until 1916 that it was discovered that the only way to make it function reliably was to open or close it only at, or very close to, the top of the tide. This seriously reduced the entrance availability, but the alternative was the risk of that it simply would not move, possibly breaking its draw cables in the process.[45] The reader may also have wondered how, in wet- dock mode, ships could tie up without risking damage to their bilges on the altar-courses. The answer is that they could not, but it was not until 28 January 1914 that the Board authorised the purchase of eight twenty by four-foot floating fenders. When these were found insufficient, a further eight were acquired in June.

Although the pumps had been installed a little late, it would not have been the Port of Liverpool if they had not then become ensnared in the characteristic rows between the Board and its customers. On 25 March 1914 the Docks and Quays Committee heard that "some" of the engines and pumps had been tested but that Thomas Newell, the new Engineer-in-Chief, was having difficulty getting and keeping the dock empty to run first a full formal trial for his own satisfaction and then later the final acceptance trial. First *Mauretania* and then *Aquitania* stayed in there for inordinately long times, and the Royal Navy entered the equation with HMS *Erin* and a number of other warships which were enigmatically recorded as "Job No. 33" etc. It was not until 26 July 1915 that the pumps could be formally signed off as satisfactory.[46]

During the time that the interim dock was "advancing rapidly towards completion," as engineers used to say, matters had largely to be left in their hands, which meant that the Board had time to think about other things. Although Gladstone had probably already undergone more changes of plan than any other dock in the world, it was decided by the Board on 22 December 1913 that instead of two 870-foot-long locks, the new entrance should have one 1070-foot-long lock. Two weeks later, the Works Committee heard Mr. Holt suggest that the shed on South Number 1 Branch should be enlarged from two stories to three. But we must not mock these gentlemen too much, for they made two changes in early 1914 which were absolutely crucial. The three-story shed was eventually adopted in June 1914, and it was specified that it was to have railway lines at the dockside. This was the first dock in Liverpool to be planned from the start with direct ship-rail access. The decision was followed by one to make crane provision of unprecedented scope and flexibility, with twelve quayside portal cranes and ten

---

[45]NML, MAL, MD&HB, Works Committee, minutes, 21 January 1916.

[46]*Ibid.*, WUP 118/1.

rooftop cranes on the landward side. The cost of £226,500 was very high for a shed, but this genuinely was a new generation of shed technology for the port.

The second crucial decision was a much quieter one, indeed it is not clear from the records exactly who made it or when, but it was alluded to in the Works Committee meeting of 3 September 1915. In early 1914 there had been difficulty getting anyone to tender for the construction of a rolling caisson for the river entrance, to the point where the Works Committee considered the possibility of inviting a German firm to bid. On 3 September it accepted the tender of Sir W.G. Armstrong , Whitworth and Co. to build three pairs of mitre gates. Given the track record of rolling caissons both earlier at the interim dock and later at Langton Entrance, it is probably fair to assume that by avoiding such problems at its new main entrance the Board did the prospects of the entire new system a great deal of good.[47]

Meanwhile, out in the wet and the mud, Newell's men were carrying on with "snagging" from the graving dock. The only obvious in-house error seems to have been a miscalculation of the amount of hydraulic power required, which was rectified by laying a main across to the Langton hydraulic pumping station at a cost of some £700.[48] As the dock and its snagging neared completion, men and machinery were gradually transferred onto the main project, and in early 1914 Newell was looking for more machinery, presumably intending to increase the already high rate of work which had gone on at the graving dock. On 4 February the Board approved the purchase of four concrete mixing machines, in March of a steam navvy and in June of fifty side-tipping railway wagons. The outbreak of war did not cause an immediate labour shortage, but as casualties mounted, conscription was introduced and ever- more battalions were raised, the problem grew rapidly.[49]

In January 1916 the Works Committee decided that the work should be ceased as soon as the underpinning of the river wall was completed. For almost exactly a year the Committee considered various proposals to employ German prisoners of war as navvies but found it impossible to get the Army to meet all the costs of guarding and accommodating the prisoners. This apparent looking of a gift horse in the mouth was nowhere near as foolish as it might seem: there was

---

[47]It is, of course, possible that the caisson had been Lyster's idea and that it took Newell a while to gain the members' trust sufficiently to dare suggest changing it.

[48]But see below for an alleged major mistake.

[49]The King's (Liverpool) Regiment had, at its peak, twenty-six battalions, not including those local men enlisted in Corps (rather than regiments), such as the Royal Engineers or the Pioneers – to whom the Dock Engineer's younger men of every description would be highly desirable recruits.

a considerable corpus of parliamentary reports and professional engineering literature on the costs of the "free" labour of English convicts, which could in fact prove extremely expensive, not least because the output of convicts rarely exceeded thirty percent of that of proper navvies and could be as low as twenty percent.[50] As the workforce continued to dwindle the question was raised again, and it continued to be discussed on and off until the meeting of 8 November 1918. At the Armistice, the works were still not completely stopped, but within the week Newell was instructed to "secure Irish or other labour" to get the works moving again.

This was not simply a matter of seeking a return to "normality," for during the war there had been frequent exchanges of acerbic communications with the Admiralty, which viewed the availability of the Graving Dock for battleships and for the large requisitioned passenger liners as a matter of great importance. Indeed, the priority they attached to this was such that one might almost have thought they had contributed to the cost of its construction, as they would have done in any other European country. It soon became clear that the dredged channel forming the temporary river entrance was (as had been expected) extremely prone to silting, and keeping it available for vessels drawing thirty-five or more feet involved frequent, extensive and costly dredging: for example, on 7 June 1918 the Works Committee heard that some 5000 tons of silt per week were being deposited. The dredging cost of a "one-off" visit by an exceptionally deep vessel, such as *Leviathan* (ex-*Vaterland*), which docked on 4 November 1918 drawing thirty-eight feet, was over £1000.

There had also been a narrow escape on 14 July 1918, when *Aquitania* was being docked and the caisson refused to close. Fortunately, the ship was over the blocks before the tide fell to the point where it might have been in danger, but this failure, together with the silt problem, meant that the dock could not be operated either cheaply or reliably until it had been enclosed within the rest of the Gladstone system. The temporary entrance had been a bold attempt at solving a particularly intractable commercial problem, but it had to be regarded as perhaps more temporary than intended. By 1921 the caisson needed scaling, painting and minor repairs, for which it was necessary to float it out to Queens Graving Dock, where it remained for many weeks, during which time the estimated cost of the works rose from £4100 to over £6000 and a considerable amount of silt settled

---

[50]In the case of penal servitude providing work for the convicts, not building a dock or harbour, was the key objective. It was almost entirely confined to government-funded harbours of refuge, because any commercial port constructing a new dock wanted it finished and earning revenue as soon as possible after expenditures began. The extreme converse case was Peterhead harbour: built by convict labour, it took almost a century.

in the temporarily tidal dock.[51] It may not have been unwelcome to the Board that the following year the works on the rest of the system reached the point where it became necessary to close the graving dock in order to build the rest of the works around it and in particular to construct the wall which would enclose the vestibule dock, cutting the graving dock off from the river.

In the immediate aftermath of the war, we again find indecision abroad. In early 1919 the port was still acutely congested, and at the meeting of the Docks & Quays Committee of 8 January it was agreed that the Engineer should have the use of the West Hornby shed (which was prime berth space) for the storage of cement, and that the works should "be pushed on with all possible despatch." The shortage of labour continued for a little while: on 24 January the Works Committee approved the action of the Engineer in taking on "all youths over the age of 18 who are likely to be able to do a full day's work at the ordinary standard rates."[52]

Labour shortages could be ameliorated by the deployment of more machinery. In September the Works Committee heard that only 484 men were at work, and in February 1920 the Engineer purchased, among other things, two extra locomotives for spoil haulage, additional piling equipment and another steam navvy. But there were shortages of materials too: the West Hornby shed was found in June 1919 to be almost empty, and it was not until October that a large amount of cement (1200 tons) arrived there. Things were gradually coming right for Newell, who had only succeeded in spending £340,534 at Gladstone in the year ended 1 July 1920: in 1922 this had risen to £523,546. The Board had by then undergone another change of heart: on 6 January 1921 it had decided not to "push" the works but to carry them out in the most economical manner. This does not mean, of course, that the expenditure could not rise – it did – but merely that Newell was not to "throw money at the job" in the interests of speed. The reason for the change is simple: the postwar bubble had already burst and congestion on the quays was no longer a problem. This meant that the opening of Gladstone could be expected to yield some efficiency savings (particularly through its very superior sheds and craneage and longer hours of entrance usability). It was not, however, likely to generate new trade for the port, only to provide better facilities for existing customers.

---

[51]The problem with the caisson was brought to the attention of the Works Committee on 19 July 1918. Such was the pressure for docking large vessels returning from naval requisition that it continued in slightly risky use until 1921 and was reported back *in situ*, but with the dock needing dredging, to the Docks and Quays Committee on 12 October 1921.

[52]It was not normal practice to take on labourers under twenty-one years of age.

In one sense, Gladstone was to prove a happy accident for the Board. Its completion was exactly the sort of job of which the Unemployment Grants Committee (UGC) approved because it was highly labour-intensive and could accommodate plenty of men who were either unskilled or who possessed skills for which there was little or no demand. The UGC was empowered to make grants only to major capital schemes "of public benefit" carried out by public or statutory bodies, and over the years of its existence (1920-1932), port improvements accounted for 10.9 percent of its expenditure. Its grants were made only towards the interest cost of projects but were often an important factor in a port authority's decision whether to go ahead or not. In the case of Gladstone the decision to go ahead to complete the project was already in place, and what the UGC provided was funding to allow it to go forward more rapidly - allowing in effect a resumption of "pushing" the works. This was crucial, for a half-dug dock was a fearful waste, costing the interest on what had already been spent, representing a need to spend more, and producing not a penny in revenue. This, and the determination of the Board to see it through, enabled completion while the state of trade was poor and safely before the Great Depression burst upon the port. The royal opening on 19 July 1927 may have lacked the splendour of some previous openings, but it was still the biggest party Liverpool had seen for years, as well it might be, for had construction lagged on into the Depression it would almost certainly have been suspended, with consequences which might be difficult to quantify but are simple to envisage.

The situation was not as bad as it might have been because at least the Board emerged from the war in a fairly strong financial position. Whereas the shipping industry seemed to spend most of the next decade looking back wistfully to the state of trade in 1913, the Board's income from dues rose from £1.58 million in 1913 to £2.1 million in 1919, and between then and the completion of Gladstone in 1927 it never fell below £2.4 million. As a result, Newell's expenditure was restricted more by labour-and-material shortages than by the availability of funds, and after the rise mentioned above, it passed the £1 million mark in each year between 1924 and 1926. Nor was this rate of work achieved at the expense of neglecting everything else: quite substantial sums were laid out on plant modernisation, the building of new hopper barges and above all the improvements to the Alfred Entrances (see chapter four). The fall in interest rates combined with the rise in income to take the percentage of dues income spent on debt charges well below fifty percent, a level it had not reached within living memory.[53]

---

[53]NML, MAL, MD&HB, *Annual Accounts*, for the years of heaviest expenditure show new bonds being issued at five percent at the same time that old ones were being paid off at six to 6.5 percent.

The Board was nevertheless living dangerously because the shipping industry in general was demonstrably depressed. The Board had followed the example sometimes shown by its predecessors by spending counter-cyclically during a recession (when labour and materials were cheap) in the confident expectation that by the time the works were completed good times would have returned. It was risky, but it had worked in the past. This time, not better but even worse times were just around the corner. Yet Gladstone had a key part to play in the Board's attempts to minimise the effects of depression when it came. It was a simple matter of efficiency: temporary closure of older entrances and berths and concentration of what traffic there was in the brand new facilities at Gladstone enabled significant operating economies to proceed hand-in-hand with providing a service for the diminished number of ships still using the port.

The issue of how well the Board performed in its strategic role is considered in a wider perspective in chapter four. In the specific case of Gladstone we can divide events into two fairly distinct periods. Between conception and the near-cessation of work during the war it behaved indecisively, and indeed its treatment of the numerous different plans it requested verged on the capricious. The clearance of the Hornby spoil bank in 1904-1905 was an act of sheer incompetence, knowing as it did that it would cause land acquisition costs to rise significantly. Yet by 1911 substantially the same men were able to act decisively and to take the important and dangerous decision that as soon as the interim dock was finished they would continue with the main scheme. In the event, they became even bolder and re-started the main scheme over a year before the graving dock was finished. This led into a period in which, as we shall see, the standard of policymaking was improved, enabling the port not only to ride out the depression and to be ready for the return of trade when it ended but also to carry out some smaller-scale developments for which the need had been conspicuous since the 1890s. Some sense of vision seems to have returned, albeit on a smaller scale than in some of the great developments of the past.

There remains one mystery. Kenneth Longbottom points out that although the Gladstone graving dock in its complete form was nominally 1050-feet long and the river lock serving it was 1070, there was not room in the vestibule dock to swing a vessel longer than about 850 feet into the graving dock.[54] He suggests that this was a well-known and recognised "blunder," which among local seafarers and port workers it may well have been. But it was not necessarily known elsewhere. The account of the opening of the complete system in *Engineering* gives the dimensions of the dock without mentioning the access problem, and E.P. Cotter's normally reliable account of the Port of Liverpool does the same, also describing it as the largest dry dock in Europe. In the

[54]K. Longbottom, *Liverpool and the Mersey. Vol. 1: Gladstone Dock and the Great Liners* (Peterborough, 1995), 52.

analytical pages of Francis du Plat-Taylor there is no mention of the access problem.[55] Stuart Mountfield remarks on the unfortunate necessity for the interim arrangement, which he admits was wasteful, but makes no mention of access in the finished scheme.

Resort to a large-scale Ordnance Survey map and a scale cardboard cut-out of a 1000-foot ship of 110-foot beam reveals Longbottom's claim as an urban myth and his small-scale diagram as inaccurate. It *is* true that a 1000-foot ship could not get into the graving dock if the berths on the west side of the vestibule dock were occupied, but what is interesting about the myth is that it could have survived. One would expect that the alleged problem would have to be publicly revealed, at the latest in 1968 when the graving dock was opened in a new role as an interim container depot pending the completion of Royal Seaforth. The fact that it was not suggests that Longbottom is wrong, but an investigation of the size of vessels using Gladstone at that time reveals that the reason for the absence of complaint or controversy was simply that it did not matter. The largest dry cargo carrier in the world in 1968 was the roll-on roll-off vessel *Admiral William M. Callaghan*, which was only 694-feet long. OCL's *Encounter Bay*, at 29,150 deadweight tons the largest container vessel afloat or under construction at the time, was 745 feet.[56] The initial (1963) plan for Seaforth had envisaged vessels of up to 800 feet. Gladstone's ability to handle 1000-foot ships was not one but three generations ahead of its time - or so it seems.

In reality, the maximum size of ship which can be handled may be a convenient figure to bandy about for advertising purposes, but it is by no means necessarily the most important attribute of a large deep entrance: Gladstone's entrance had about forty feet of water on the sill at mean high water, which meant that the largest vessels then using the port could never get "neaped" and that more modest vessels could lock in and out at most states of the tide. It was that availability which both contributed to a quicker turn-round at Gladstone and allowed the problematical Canada Basin and entrances to be temporarily closed down as an economy measure during the Depression.[57]

---

[55]E.P. Cotter, *The Port of Liverpool, Including Birkenhead and Garston* (Washington, DC, 1929); F.M. du-Plat Taylor, *The Design, Construction and Maintenance of Docks, Wharves and Piers* (London, 1928); and S. Mountfield, *Western Gateway: A History of the Mersey Docks and Harbour Board* (Liverpool, 1965).

[56]*Fairplay,* 13 June 1968, 28.

[57]This point is strongly made in L.A.P. Warner, "The Mersey Dock Estate - Its Management and Principal Trades," *Journal of the Institute of Transport,* Special Liverpool Congress Number (1928), 443-460.

Gladstone also stood the test of time remarkably well. During the period 1840-1900 new docks had typically only lasted for about twenty years before they needed either major updating (often involving technically difficult and hence costly deepening) or replacement by another generation of new dock. The general concept of Gladstone was more or less finalised by 1913, when the Board took the final decision that the entrance lock should be built for 1000-foot vessels at an estimated additional cost of £118,000. It remained possible to make other alterations of considerable importance, such as the ratio of shed space to length of quay, but even those changes had to cease by 1925 as the drive to completion began. The decision to build all the transit sheds to three storys despite the heavy cost was important because while the size of ships did not rise as much as expected, the cargo density (tonnage of goods per foot length of vessel) continued to increase, and that is what determines the amount of shed space needed per yard of quay. This reflects Newell's belief that Liverpool was unlikely to have to accommodate express passenger liners, which had much lower cargo densities than either intermediate passenger or cargo liners. As early as 1921 he stated that he "doubted if any more *Aquitanias* or *Olympics* would be built."[58] He was, of course wrong, but their successors did not sail from Liverpool. In 1975, when the break-bulk cargoes it was designed to handle were fast disappearing, the dock was still able to accommodate any size of ship people might wish to put in it, and its equipment, while no longer "state of the art," was still good.

It might be suggested that the last few paragraphs serve only to show that Gladstone was too big, too durable and too expensive. They certainly do not prove that the Board necessarily exhibited, as is sometimes assumed, a remarkable degree of foresight, for the 1000-foot ships, when they eventually did arrive, would not be break-bulkers wanting to use Gladstone.[59] There were, perhaps,

---

[58]In discussion of H.J. Deane, "The Development of Cranage Facilities for Discharging Vessels of the Largest Size," *Proceedings of the Institution of Civil Engineers Conference (*1921), section V, 21-24.

[59]It is highly doubtful whether a 1000-foot break-bulk vessel, which would have a cargo density of about seventy tons per foot length, or about double that of a good modern 600-foot cargo liner of the 1920s, could under any circumstances have been an economic proposition. It would probably not have been possible to shift the cargo fast enough to prevent its stays in port being about double as well. Its draft, about sixty feet, would exclude it from almost every port in the world. The idea had a good deal of currency, however, thanks in part to Sir John Biles, who approached the problem as one purely of the cost of actual sea transport. He was several times ridiculed for supposedly advocating such vast vessels. In fact, he willingly admitted that it was a theoretical notion and the port situation might render it impossible. See J.H. Biles, "The Draught and Dimensions of the Most Economical Ship" *Transactions of the Institution of Naval Architects*, LXXIII (1931), 243-263.

elements of luck in Gladstone's success, but what was needed was a long-term solution: the problems with other parts of the Dock Estate, including the pressing need for major improvements in facilities for the coasting trade, the re-construction of Alfred Entrance and the ongoing electrification programme meant that it was essential that Gladstone worked the first time - and for a long time. The major elements of it were successful, and that was in part because what resembled a continuation of earlier indecisiveness was in fact an ability of the Board to think on its feet and allow the project to evolve to meet a changing perception of what it was there for.[60]

[60]The strange thing in all this that the membership of the Board was relatively unchanging - and unpromising - but that, too, is an issue for chapter four.

Chapter 3
Not Before Time: The Board Keeps Its Promise

Thus far, the concern has been with the handling of large and/or exceptional ships, with the need to make large investments to keep up both with the growth of these vessels and with the ever-increasing quantities of goods they sought to discharge and to load with ever-faster turnaround. But these were not the only ships to use the port, and we must now turn our attention to the provision of facilities for more humble craft.

In the earliest days of the Dock Estate, docks were built for the "big" ships, and the coastwise trade was left more or less to chance. The first dock to change that was Clarence, opened in 1830 specifically for steamers, which at that date were all in the coastwise or short-sea trades, and it was followed by a handful of other docks designed for coasters, of which Collingwood (opened 1848) was the last. After that, coasters had to survive as their ancestors had, on facilities handed down as outgrown by the trades using larger vessels. The allocation of a dock to the use of coasters thus came to be a matter not of what it *was* suitable for but rather of whether it was suitable for anything better. This repeatedly emerges in discussions at the Board and in print:

> some of the Port Authorities have treated the coastwise trade as the Cinderella of the transport world, and the failure to meet its requirements, and especially its relegation to the outworn and obsolete sections of their undertakings has added enormously to its handicap in the competition with railway transport.[1]

This only meant, however, that accommodation for the coasters would probably, although not necessarily, be unsuitable. Nelson Dock, for example, was handed down to the larger class of "Irish Boat" and proved extremely satisfactory. It had been built for the North Atlantic trade, opened in 1848, but it was sufficiently deep and spacious for the largest coasters of the early 1900s, had reasonably wide quay margins and was connected to the railways.[2] But Nelson was a happy

[1]Anon., *The Coastwise Trade of the United Kingdom, Past and Present, and Its Possibilities* (London, 1925), 64.

[2]There are, however, grounds for suspecting that Nelson was built with the intention of eventual "down-grading:" Wellington was in hand and Canada and Huskisson at "concept stage" before it was completed.

example: many of the other docks reserved for coasters were apparently unsuitable for almost anything. The prime example was the Clarence "system," consisting of a wet, half-tide and two graving docks, as well as a graving dock basin with a graving grid. They were all small - Clarence Dock itself was much the largest water area at just over six acres and a sill just three feet three inches below Old Dock Sill (ODS), which gave about eighteen feet of water at mean high water. The sheds were original equipment (albeit slightly extended) and therefore of insufficient width to accommodate the cargo of a modern coastwise steamer, and the avenues behind the sheds were too narrow to allow the construction of wider sheds. There was no hydraulic power for gates or lifting appliances (which amounted to one twenty-five-ton hand crane), and despite being very close to the main spine of the dock railway, it was not connected.[3] The gas lighting in the sheds and on the quays was pitiable. About 120 feet of its quay space was given over to the Engineer's Rubbish Depot. No money beyond routine maintenance had been spent on it since 1830, and anything of remotely comparable obsolescence which survived nowadays would be "listed" and turned into a "maritime heritage experience."

If we neglect appearances and consider instead the tonnage handled per lineal yard of quay, however, Clarence suddenly looks well short of useless. In 1905 the biggest dock in the system was Alexandra, which handled 1,204,620 tons, a rate of 307.22 tons per yard of quay; Clarence, for all its age, still handled 188,499 tons, or 121.69 per yard of quay.[4] The entire Birkenhead system, so praised for its deepwater quays and rail connections, actually handled less (110.56) tons/yard than the obsolete Clarence, a comparison which is the more dramatic because while Clarence was mostly handling general cargo tramps, Birkenhead had the tonnage-boosters of coal export and bulk grain imports. The best dock in the coasting trade, however, was Nelson, which turned 477,779 tons at the astonishing rate of 595.02 tons/yard, second only to the recently-modernised top dock, Huskisson, with 656 tons/yard. Hornby, Toxteth and Harrington, the latest *de novo* ocean-going docks in the system, all fell well short of Nelson in tonnage per yard. This might seem to indicate that extensive investment in

---

[3]E.P. Cotter, *The Port of Liverpool, including Birkenhead and Garston* (Washington, DC, 1929), provides useful tabulated data on the facilities, or absence thereof, at individual docks.

[4]These and similar figures are calculated from the National Museums Liverpool (NML), Maritime Archives and Library (MAL), Mersey Docks and Harbour Board (MD&HB), Revenue Statements, various years, for tonnages and dues, and from the the *Yearbook*, various years, for lineal quayage measurements. Clarence's efficiency is actually a little over 125 tons per yard if the rubbish depot is subtracted.

facilities for the coastwise trade should have been a priority target. The reason why it was not was a simple matter of money.

Figure 3
Transferring Bulk Grain to Barges at Alexandra Using a Floating Elevator

Source: National Museums Liverpool, Photography Department, Ref. No. N86-283.

Although the Board was originally established to act as provider of a "public good"[5] – that is to say it was to spend its dues income disinterestedly for the benefit of the trade of the port – it soon became apparent that it would have to behave like a business. If it was to be able to keep abreast of the constant changes in ships and trades, which were requiring it to make ever-larger investments, it had to seek a return on its capital. To the decision-makers it was

[5]Under the terms of the Mersey Docks and Harbour Board Act, 1857, section 56.

not so much volume per yard (though they liked volume as a figure to emphasise the importance of their port) that counted as dues income per yard. Here the various dues concessions which applied to the coastwise trade make it and the docks serving it look less impressive. Clarence, for example, earned only £8713 at a rate of £5.62/yard. Nelson's volume translates into a value of £23.34 per yard, which is surprisingly good compared with Alexandra's £36.01, but less good when compared with Harrington's £57.05/yard or Huskisson's £71.47.

The gentlemen of the coastal trade were in a line of business in which they could make a fair profit, but probably only by using port facilities that came very cheap.[6] They repeatedly claimed that if they were unable to pay their true share of the capital costs of the port's major works undertaken for the larger vessels. This may be true, but it should also be remembered that it was the cargoes of the large vessels which produced such a volume of work for coasters as to turn Liverpool into a "honey pot." The humble vessels using the cheap facilities in Clarence were also using the expensive facilities of the deep-water docks, albeit only indirectly. Furthermore, the dredging of the bar and main channels, undertaken at huge expense for the large vessels, had the side-effect of allowing coasters to come and go as they pleased, even at low water, spring tide (LWST); only docking and undocking were affected by the state of the tide. It seems fairly clear that the most important things for coaster owners were generous entrances (to allow flexibility in sailing times) and wide sheds and/or quay margins to allow rapid discharge and loading. The absence of power-driven cranes was less important: these people were used to using their own tackle.

Unlike the owners, the Board was seeking only to break even. This is the fundamental source of the long-standing rows which soured the trading atmosphere of the 1890s and beyond.[7] The trade understandably wanted better facilities while the Board, equally understandably, thought it was already doing favours enough for the trade by letting them have adequate, if modest, facilities at relatively give-away prices. For a majority of the Board to agree to spending significant money on improving facilities for coasters, some other factor or factors in addition to the constant complaints of the coastwise owners, had to come into play. Furthermore, it had to be a powerful one, for such a heavyweight

---

[6]Ports which dealt mainly in general coastwise trade, such as Connah's Quay, tended to have very rudimentary facilities. This, of course, does not apply to specialist coastwise mineral ports such as Maryport.

[7]Some of this rancour is brought out in A. Jarvis, "The Port of Liverpool and the Shipowners," *The Great Circle*, XVI (1994), 1-22.

figure as Alfred Holt had already taken up the cause of the coasters as an essential distributive ingredient of the port's overall trade and had been voted down.[8]

One such factor did arise. From the pioneering days of the Liverpool and Manchester Railway onwards, the role of railway companies in the politics of the port, and sometimes of the city as well, was rather like that of a pantomime villain, on whose foul machinations almost every ill could be blamed. As Jack Simmons has shown, railway companies were not only very large and powerful organisations but could also be remarkably inept at "customer relations."[9] During the last third of the nineteenth century it became public knowledge that railway companies did not seek only short-term profit, and two implications of that policy greatly enhanced the suspicion - and fear - with which independent port authorities viewed them.[10]

The first hinged on the question of unfair rates, as defined in the Railway and Canal Traffic Act of 1854, and what we now call "predatory pricing." Examples of this included charging high rates on routes where there was little or no competition in order to subsidise unprofitable rates on routes where there was. Freight rates from Hull to Leeds were much lower than rates from Liverpool to Leeds because the Aire and Calder Navigation was a large, efficient and independently-owned waterway providing effective competition. Rates between Liverpool and Manchester were astronomical because the three railways offering services operated a price-fixing agreement and also controlled the Bridgewater Canal. The resulting local surplus was available to subsidise other routes.[11]

---

[8]As, for example, at a meeting of the Works Committee, 21 February 1890, when it was "Resolved (Mr. Holt dissenting) that the projects for the improvement of the docks between the Nelson and the Waterloo docks be not entertained." This issue can be followed in NML, MAL, MD&HB, Worked-up Papers (WUP) 17/1.

[9]J. Simmons, *The Victorian Railway* (London, 1995), 261-264.

[10]This was already evident among witnesses before Great Britain, Parliament, *Parliamentary Papers* (*BPP*), Royal Commission on Railways, 1867 (3844), XXXVIII, part 1, 32. It is noteworthy that, much later, when the Dock and Harbour Authorities Association was established, it excluded the port operations of railway companies from membership.

[11]Examples of unfair rates were regularly paraded before Parliamentary Committees, but for a massive collection of specifically Liverpool complaints see the pamphlet *Report of the Meeting of the Deputation from the Joint Committee of the City Council, Dock Board and Mercantile Associations with the Directors of the Railway Companies Who have Termini in Liverpool* (Liverpool, 1881). Browsing the hand list of NML, MAL, MD&HB, Legal C, will provide boxes full of similar material.

The second grievance independent port authorities had against the railway companies was their ownership and operation of ports. Associated British Ports is still the largest port operator in Britain, and its ports are the inheritance of the former British Transport Docks Board, which had been established to run the former railway docks when the railways were nationalised. It was openly admitted that a surplus like that in the Liverpool-Manchester trade might be deployed in subsidising services to Fleetwood or Holyhead in competition with Liverpool (and later Manchester as well) in order to increase the overall traffic on their system. In the case of Liverpool it went further than that: not only did the London and North Western Railway (LNWR) invest heavily at both Garston and Ellesmere Port, but both the Cheshire Lines Committee (CLC) and the Great Western Railway (GWR) attempted to build rival facilities right on Liverpool's doorstep at Otterspool and Tranmere, respectively. The Board succeeded in thwarting the GWR scheme in Parliament, while the CLC scheme did get built, but it was a complete failure.

These background grievances were exacerbated by the continuing tendency of railways to amalgamate into ever-larger companies. The Board had fought hard to prevent the proposed amalgamation of the LNWR with the Lancashire and Yorkshire in 1871, which would have left Liverpool with no direct connection to its hinterland independent of the new company.[12] Between 1880 and 1914 Liverpool as a whole (the Board, City Council and Chamber of Commerce) examined a number of schemes, including Alfred Holt's Plateway, several proposals for building a new freight-only railway to Manchester, and even one for carrying freight on municipal tramways at night.[13] In 1899-1901 there were the famous Liverpool Steam [road] Vehicle trials. All these initiatives aimed to harm the traffic being carried in and out of the port at premium rates by the existing main line railway companies, thus damaging the flow of funds available to subsidise Fleetwood, Barrow and the rest.

By 1905 a few members of the Board were already coming to the view that nurturing the coastwise trade was another way of hurting the railways, a shift not unconnected with the election of Alfred Read of F.H. Powell and Co. in 1904. The investment in wider quaysides at Princes Dock, by now used almost exclusively by the coasting trade, paid off with a distinct rise in both tonnage and

---

[12]For detail, NML, MAL, MD&HB, Legal C6. See also J.R. Kellett, *The Impact of Railways on Victorian Cities* (London, 1969), 188-195.

[13]This apparently foolish idea continued to attract serious consideration. In 1913 a report on it was published by the General Manager of Liverpool Corporation Tramways. NML, MAL, MD&HB, P/2/13/4. For the rest, see Liverpool Records Office, 380.COM.

revenue.[14] This was the more gratifying for being in defiance of the relative stagnation in trade as a whole during the "Edwardian Depression."[15] We must, of course, question whether the investment was worthwhile, which proves not to be the simple question one might expect. Tonnage and revenue both rose by about twenty percent, some £2000 per annum, and the only expenditure on the improvements which appears in the *Annual Accounts* is £23,150 "on account" from the Unappropriated Receipts Account in 1905, but this sum was transferred to capital account the following year and "disappears" into a very large pot - along with any further expenditure on the project. As a result, we cannot tell whether the £2000 represents a fair return. The construction of Riverside Station (opened to passengers in 1896) meant that there was already a rail connection available on the west side of the dock, though the only way it was utilised at this stage was via a "back door" to the passenger station.

It might be thought that these improvements were irrelevant in railway terms because most of Princes' customers were in the Irish trade and trains do not float. This, of course, is not a valid objection, for Princes was in competition for the Irish trade with railway-owned ports, chiefly Holyhead, but also in varying degrees with Fishguard, Heysham, Barrow and Fleetwood. Integrated systems such as the railway companies built up were a source of great strength in some respects but had the weakness that a concerted attack at any one point might damage the entire system.

Other things being equal, customers were likely to view railway ports with disfavour. The last third of the nineteenth century was marked by literally dozens of parliamentary enquiries and reports bearing directly or indirectly on the railway industry, and they provide a backdrop of remarkable hostility to railway companies, which were attacked for, *inter alia,* unethical financial practices, unsafe working practices, unfair pricing policies and suppression of competition. Malicious jokes about railways were current well over a century before the

[14]This chiefly consisted of a ferro-concrete staging built out into the water to provide a wider quay margin.

[15]The quotation marks are to draw attention to the fact that in Liverpool both tonnage and revenue increased considerably between 1901 and 1910: in port terms the "depression" was in fact a couple of minor dips in the graph and a general slackening of growth. The reason for this "escape" is unclear but is probably the same one which enabled the port to ride out the financial storms of 1847 and 1866, namely the great diversity of its trade. The town at large did less well; see, for example, M. Simey, *The Disinherited Society. A Personal View of Social Responsibility in Liverpool during the Twentieth Century* (Liverpool, 1996).

government gave Railtrack to the humourists.[16] The Royal Commission on Canals heard probably as much evidence on the iniquities of railway companies as on the benefits of canal investment. During the Edwardian period, pressure was again building for the nationalisation of the railways, coming to a head with a Select Committee which heard evidence in 1912.[17]

This might be thought to form the answer to the Dock Board's prayer: railways would become a public good, just like the Port of Liverpool, and the two organisations would collaborate to provide a better service at a lower price. The Board saw matters quite differently, and William Thorne, its solicitor, gave evidence firmly against the bill.[18] The reasoning was simple, but it also indicated the extraordinarily deep-rooted nature of the Board's mistrust. At present, although the situation was undesirable, there was a limited amount of competition between the different railway companies using the port; while far short of optimal, it was better than nothing. If the railways were nationalised there would be no competition at all, and the Board simply did not believe that government would be able to resist the temptation to make their operation look efficient by employing similar tactics to those of the existing companies. They probably also believed that key top managers would be retained and "once a railway manager, always a railway manager."

Matters arguably got worse before they got better: the Second Report of the 1918 Select Committee on Transport concluded that "in this country it is unquestionable that all other forms of internal transport are subordinate and ancillary to the railway system" and that "it may be open to discussion whether the railways should be encouraged to extend their ownership of ports." This suggests a clear presumption in favour of further railway amalgamation.

Although the 1921 Railways Act required that railways carry on their ports business "as far as possible on a remunerative basis and not to support them out of earnings from the railway proper," the 1929 Royal Commission on Transport heard that this was not observed.[19] Furthermore, in the evidence of

---

[16]The one about the man who laid down on the Midland Railway main line to commit suicide "...poor devil died of starvation!" was re-told as old in 1880. The sandwich-curling machine in railway buffets had long been an essential part of the corpus.

[17]There had been previous waves of hatred of railway companies which manifested themselves in calls for nationalisation in the late 1840s and the mid-1860s.

[18]NML, MAL, MD&HB, Legal C65.

[19]This was, of course, the famous "Grouping Act." The majority of non-shipping commentators scarcely noticed the coasting issue at all. The Board had, however, enjoyed some minor successes at the report stage, helping to secure the "non-grouping" of both the Cheshire Lines Committee and the LNWR/GWR Joint route from Chester to Birkenhead,

Earnest Foley of the Board of Trade, it heard chapter and verse on the alleged railway attack on the coasting trade. Foley stated, *inter alia*, that railway pricing policies were "unreasonable, and disadvantageous to the country as a whole." The railway industry's view was, of course, that it only wished it could charge higher freight rates; the "special rates" were a regrettable device needed to prevent loss of custom. Worse, the road transport industry was now joining in, when "the coastwise trade is, or should be, much cheaper than either [road or rail] of these." The positive side of his evidence included statements that "coastal shippers must not, therefore, be allowed to be crippled" and "money spent on facilities for coasters would be to the public advantage." This, however, only confirmed the implications of the evidence of Sir Herbert Walker, General Manager of the London & South Western Railway, to the effect that "[a] railway company has two objectives in providing docks, the dock proposition being the lesser by a long way. Its main object is to create a traffic to carry over its rails."[20]

This contrasted sharply with the evidence of Sir Norman Hill, who argued the case for government subsidies for port development because "the cost at which such services are provided are matters which affect every man, woman and child in the land." There followed a ferocious attack on various railway practises which do not need to be enumerated again. The findings of the Commission mark a realisation of what the railway companies were up to, or rather of what the shipping men said they were up to, and support the notion that previous attempts at modifying their behaviour had been unsuccessful. There is a definite sense of a change in the climate, perhaps even a partial recognition that:

> During the sixty years that have elapsed since 1865, there has been a steadily increasing wastage of [British] sea power generally, due to the failure of the nation and its governors to recognise the true value and importance of British seaports and the necessity for maintenance of a regular service of ships operating between them.[21]

There was, of course, another side to this long-running clash of interests. The railway companies could claim with some justice that they had massive arrears

---

thus severely denting the local monopoly which the LMS would otherwise have exercised. *Ibid.*, Legal C64, C65.

[20]*BPP*, 1918 (130,136) IV.465, Select Committee on Transport, q. 718, *et seq*.

[21]Anon., *Coastwise Trade*, 56. Note that this statement can only be justified in terms of unfulfilled aspirations: if it is taken to refer to actual shipping tonnages then it is true only for the post-1913 period.

of maintenance to overcome at the end of the war, amounting to some tens of millions of pounds. All their costs had risen, and some, notably wages and coal, had escalated dramatically. Railway shares had, in the aggregate, suffered large price falls, making it necessary for companies to pay high dividends in order to prevent complete collapse. But the purpose of this chapter is not (fortunately) to measure the rights and wrongs in the long-running squabble between railways and coastal shipping but merely to explain the MD&HB's change of heart in favour of the coasters. The answer seems to lie in a combination of the Board's long-standing dislike of railway companies with a considerable "charm offensive" on the part of the coasting trade.

In 1928 the basis of a contract for the sale of Clarence Dock to the Liverpool Corporation was worked out; it was to be the site for one of the new breed of "Electricity Super-power Stations." This provided the Board with a considerable and very welcome financial windfall but at the expense of infilling the busiest part of the old Clarence system to the considerable inconvenience of the smaller coastwise operations. But Clarence was no longer the home only of small operators, many of whom had cast in their lot with Coast Lines Limited, whose Chairman was Sir Alfred Read. Born in 1871, Read had enjoyed a meteoric rise in the coasting trade, which was reflected in his election to the MD&HB at the unusually early age of thirty-three.[22] At the outbreak of war he was Chairman of the important Bacon, Powell and Hough group of companies, and when it and others came together in 1917 to form Coast Lines, the largest coastwise group in the country, he remained Chairman. Between 1916 and 1920 he served as Director of Home Trade Services, for which he was knighted in 1919. He moved to London in 1920, which forced him to resign from the Board, but he remained a persuasive and persistent advocate of the interests of the trade.[23]

Almost as soon as the scale of the postwar recession became clear, he had begun badgering the Board for facilities which would make the coasters more competitive. In 1921 he was asking for the construction of a half-tide dock specifically for the coasting trade in the Central Docks. In 1923 he was asking for a deep-water lock to serve the same Central Docks.[24] The Board's quite understandable response was that it could not take on another major project until

[22]For an analysis of the ages of members, see chapter four.

[23]Biographical details are from NML, MAL, MD&HB, Antecedents A212.

[24]This was not because existing entrances could not accommodate the vessels, which they could, but to allow coming and going at any state of the tide, as enabled by the dredging of the channels.

Gladstone was complete and open.[25] It also pointed out that the opening of Gladstone would free up berths further to the north and might result in the coasting trade moving northwards quite voluntarily.[26] Later in the year he was back again, this time wanting the passage between Princes and Princes Half-Tide widened to enable the new Belfast boats under construction to use it safely. The building of vessels incapable of using the berths allocated to them was a tried and tested means of blackmailing port authorities, one at which Cunard were peculiarly adroit. The proposal was again turned down on cost grounds, and the possibility of the Belfast boats using berths at Alexandra was mentioned.[27] In 1924 Sir Alfred had been offered berths at Brunswick, the use of which he thought would be injurious to the coasting trades' business, and again he lobbied for the modernisation of the Central Docks.[28]

These early requests were really rather forlorn hopes, for the Board simply did not have the money to carry out the works requested. But once Gladstone was opened the situation improved. The Clarence sale eventually provided £339,744 (paid in early 1929); the Board had enjoyed a significant reduction (of some £227,000 per year) in local authority rates on its premises; and the opening of Gladstone would not only release the Engineer's Department to work on other things but also throw men out of work, making the Unemployment Grants Committee (UGC) likely to support new projects. Sir Alfred was clearly aware of these future possibilities, as there is a marked intensification in his lobbying during 1927, immediately before and after the opening of Gladstone. In February he again met with the Docks and Quays Committee, which instructed the Engineer to provide a report and plans for a new deep-water lock and a fixed jetty in the river to the north of West Waterloo.[29]

---

[25]They had, after all, sunk many millions into the construction of Gladstone, on which there would be not a penny return until it was open for business.

[26]NML, MAL, MD&HB, Docks and Quays Committee, minutes, 21 March 1923.

[27]Alexandra was probably not a sensible suggestion: it was quite a long ride on the Overhead Railway, and the popularity of the aged facilities at Princes had always lain in their convenient location in the City Centre. *Ibid.*, 3 October 1923.

[28]Brunswick had high quays and was rather distant not only from the city centre but also from the rest of the coastwise berths and the trades that were most likely to discharge over the side to coasters.

[29]NML, MAL, MD&HB, Docks and Quays Committee, minutes, 23 February 1927. A month later it was agreed that the jetty could be dispensed with. The proposal is current again at the time of writing, but the issue now is the saving of docking time.

In August he was back again with the request to widen Princes Passage, and this time the Committee seems to have been much more accommodating, giving serious consideration to widening the passage from forty-five to sixty feet and lowering the sill by two feet one inch to take it to the same level as that of the river entrance. The key concession, however, came in December, when he was offered the use of a berth at Huskisson No. 2, in among the ocean-going passenger vessels, for the new Belfast boats. With this offer came a firm promise that if the arrangement proved unsuitable, the widening of Princes Passage "would be taken into immediate consideration."[30]

It was nearly fifty years since the Board had promised the Irish trade that after the Canada-Huskisson improvements were completed, the next major work to be undertaken would be for the benefit of the coastal trade. The promise had been broken with the works under the 1898 Act, and the Board had jumped up and down on its fragments with the Gladstone project. Now, at last, things were beginning to happen, and once started, they moved surprisingly quickly.

The meeting of the Works Committee on 22 June 1928 received preliminary plans from the Engineer for extensive improvements, and these were near enough the mark for him to be told to come back with estimates and further details. It is clear that somebody was behind him with the metaphorical cattle-prod, for he produced these within the week. There were three options for the "main scheme" at estimated costs of £379,000, £694,000 and £844,000. Cutting down the "haunches" of the passage between West Waterloo and Princes Half-tide would be £2000; and enlarging and deepening the Princes Passage, £58,000. Modernising the East Quay of Princes by building a completely new shed standing on piles over the water would be £132,000 and re-constructing the graving dock to take vessels of 400 feet another £163,000. The Committee turned down the graving dock scheme and the most expensive version of the main scheme, but instructed that planning continue with all the others. There now followed a relative lull.

The Board had come to the point where it was willing to commit some £200,000 to the Princes Scheme with or without the support of the UGC, and work began in early 1929, with expenditure by July of £18,744. While this was going on, the negotiations for the main scheme continued, and a crucial Board meeting was held on 1 August 1929 at which it was agreed, after considerable discussion, that six schemes be put forward to the UGC, including the "Extension to the West Float" (i.e., Bidston Dock) at £780,300 and the Clarence improvements at £711,540.[31] This decision shows an amazing degree of confidence; remember that the UGC did not make capital grants: all it paid was a significant

---

[30]*Ibid.*, 7 December 1927.

[31]*Ibid.*, Discussions at Board, 1 August 1929.

proportion of the interest costs for a maximum of fifteen years. The Board was committing itself to additional borrowing (including that for Princes) of nearly £2,000,000 at a time when over sixty percent of its dues income was already spent on interest.[32]

On 31 October the Chairman reported that the UGC had accepted one of the minor schemes, and within a week the General Manager (Sir Lionel Warner) was instructed to "take the necessary steps to complete arrangements" with the UGC. The major schemes followed shortly with a similar instruction referring to both of them given on 29 November. The estimated cost at Clarence had by now crept up to £985,128,[33] and the UGC contribution was to be 100 percent of the interest cost until the works were completed and fifty percent for the balance of fifteen years. Work began on 22 March 1930 with the somewhat leisurely expenditure of £18,844 by July.[34]

From this one might expect that the Engineer was going to drag the job out in order to maximise the period in which the UGC would pay the full interest cost, but one would have a nasty suspicious mind, for the following year he spent £271,977, with work proceeding throughout the extensive site from Collingwood to Waterloo. From the entries in his accounts for plant (£30,848) and temporary railways (£11,186), as well as from photographs of the works in progress, it is clear that his brief was not primarily to give shovels to as many unemployed labourers as possible and see what happened; this was a high-efficiency job. On 22 April 1932, the new rolling lift bridge at the Salisbury-Trafalgar passage was opened and that at West Waterloo/Trafalgar was in place. The modernisation of the graving docks was complete, and the only major task remaining was completion of the Salisbury-Trafalgar passage. Expenditure that year was £360,984, and work was completed on 19 April 1933 at a cost of a further £80,514. This makes a "visible" total of only £732,319, a saving on the estimate of over £250,000, over one-quarter of the total.

No "run-on" expenditure in following years has been detected, and while there are some rather curious-looking credit items in the accounts they relate only to a few tens of thousands of pounds. It transpires that the estimate had been unusually generous, including large provision for the construction of "special

---

[32]"New money" at this time cost fractionally under five percent. As elsewhere in this book, the interest is expressed as a percentage of dues income only, on grounds of the relative inelasticity of such other sources as property rentals.

[33]This figure is from NML, MAL, MD&HB, Discussions at Board, 19 November 1929, but other figures were bandied about, including £973,000 mentioned in the Chairman's Report for the year to July 1930.

[34]*Ibid.*, *Engineer's Report*, 1930.

dams" which proved not to be necessary and a *further* ten percent for general contingencies. We may also presume that the project benefited from the general fall in prices and wages over the period of construction, but is still clear that the engineering management skills employed were simply in a different league from those on some previous projects.[35] As a very simple test we may look across the river to the Bidston Dock, proceeding almost exactly in parallel, with literally only a couple of weeks' difference in start and finish times. Here the project had been scaled down from £913,188 to £708,675 by the simple expedient of reducing the length of the new dock by one-third, which would clearly entail some waste of preparatory work, but even so, the works were completed on schedule and within budget, although only by some £9000. Given the very unpleasant ground conditions at Bidston, that, too, may be taken as a considerable achievement.

Figure 4
Entrance to Clarence Dock – Door to Modernisation

Source: National Museums Liverpool, Photography Department, Ref. No. N90-309.

---

[35]For a discussion of whether these skills were being misapplied, see chapter four.

The Clarence modernisation seems to have been an excellent piece of engineering. It took a collection of old docks of demonstrable inadequacy and turned them into a coherent "spine and branch" system with a reasonable sized half-tide entrance at each end. Extensive new shedding was provided (in addition to that at Princes) of an appropriate standard for an ocean-going berth; hydraulic power was applied to the old bridges and to the passage gates; and the new bridges were electrically driven. The Clarence Graving Docks were modernised, and the Princes Graving Dock was equipped with new pumps, so that vessels requiring docking could visit whichever was nearer, with the minimum travel from their cargo berth.

Was it worth it? The accounts for 1934 show the beginnings of recovery from the Depression, with coastwise tonnage up four percent as against an ocean-going increase of seven percent, while the following year the growth figures were effectively identical, within a whisker of two percent. In 1936 the corresponding figures were 2.5 and 1.5 percent. There was clearly no spectacular growth, but at least decline has been arrested for the time being. The revenue provided a less encouraging picture: the total increase in coastwise trade revenue was a mere £9436 for the whole of the port, and the coastwise dock rates inwards showed a fall of £14. The interest cost of the Clarence works alone (and not taking account of any discounting of bond issues) would be about £36,000 per annum. Despite the unfavourable economic circumstances, it was an expression of faith in the future which should eventually be justified.[36] Alas, it was not so: the new record tonnage year of 1938 showed the coastwise tonnage handled and revenue generated had increased only by about five percent, nowhere approaching the interest cost of the new works.

This suggests that the scheme, however meritorious, was poor value for money. Yet inside a year of its completion, Sir Alfred Read was back, this time pressing for the deep-water entrance to the Central Docks mentioned above. This was estimated to cost at least £500,000, and the Docks and Quays Committee decided that no action be taken.[37] One cannot help feeling that after the expenditure of nearly £1,000,000 (including the works at Princes), he was being a little ungrateful in crying for more so soon. He certainly could not plead ignorance of the Board's financial problems, for he had been a member during the anxious years of the struggle to fund Gladstone between 1904 and 1910. It is clear with hindsight that the coastwise general trade was beginning a terminal

---

[36]As suggested in S. Mountfield, *Western Gateway: A History of the Mersey Docks and Harbour Board* (Liverpool, 1965), 156.

[37]NML, MAL, MD&HB, Docks and Quays Committee, minutes, 17 February 1934; and WUP 180 (Coast Lines Berth papers) exhibit the usual bickering after as well as before the modernisation.

decline. There were resurgences from time to time, but the power of the railway companies was sufficient to allow them to do again what they had done to the canal companies, despite long-term government attempts to prevent them.[38] The rot may have been exacerbated by the poor facilities available in Liverpool and other major ports in the period 1880-1920, but by the time of Liverpool's investment it was probably too far advanced to stop.[39]

There is, however, an epilogue to the story which calls this assumption into question: in 1937 the Board relented and agreed to build a new entrance lock of very generous size for coasters at West Waterloo; what mattered was not its size but its availability to vessels drawing up to seventeen feet at any state of the tide. This was a particular advantage to the coastwise liner trade, helping vessels work to tighter timetables and thus achieve quicker turnaround. Its completion was long delayed, first by engineering problems, then by the war and the crippling amount of re-construction needed on the dock estate as a result of the May blitz of 1941, and finally by the predictable arrears of maintenance. It was finally opened by Princess Elizabeth and the Duke of Edinburgh on 29 March 1949, the same year that the re-construction of the fire-damaged warehouse at Gladstone No. 1 was completed. Then came the postwar revival in the coastal trade, peaking as late as 1962. What part did the Waterloo lock play in that revival?

The answer, sadly, is not much. The statistical evidence is, for such a recent period, surprisingly slight and inconsistent, but such of it as there is suggests that the growth of trade in the Central Docks was almost entirely confined to the "Irish boats" using Princes. In November 1947, for example, 256 vessels of 119,128 net register tons (nrt) discharged 123,317 tons of goods and outwards 224 vessels of 103,378 nrt loaded 93,092 tons of goods. The corresponding figures for November 1950, when the new lock was fully in use, are inwards 299 ships, (197,416 nrt) and 98,769 tons discharged and outwards 330 ships (181,834 nrt) and 132,936 tons loaded.[40] The large discrepancy between inward and outward goods does not occur in the month of July for the same years, and for November 1947, though not for 1950, there is a dock-by-dock

---

[38]The railway "Grouping," by establishing regional monopolies, potentially increased the power of the railway industry to harm the coastwise trade as well as the ports (see above). The GWR and the LNER, for example, each controlled virtually all the ports serving an extensive coal-exporting area in South Wales and Northeast England, respectively. But note that the LMS did not achieve a monopoly in Liverpool.

[39]Anon., *Coastwise Trade*, 64, states, for example that "this amazing declension [sic] in Liverpool's coastwise trade is partly the result of long years of neglect by the Port Authority and partly the result of grossly unfair railway competition."

[40]NML, MAL, MD&HB, Statistical Returns for months mentioned.

breakdown of the tonnages shipped which shows Bramley-Moore handling well over double the tonnage of the next-busiest dock (which was Princes). From this we may take it that a seasonal coal traffic operating from an obsolescent ocean-going facility, namely the high level coal railway, still formed a noticeable part of the coastwise trade and that the expensive new developments had not brought about any radical improvement in the coastwise liner trade in general goods. In both July and November 1950 roughly one-third of inward goods and three-quarters of outward goods were "Cross Channel Traffic."

Owing to the inconsistency of the statistical files available it is difficult to calculate exactly the tonnage per yard of quay for the impressive-looking Trafalgar Dock, intended for the top coastwise cargo liners. There is, however, no need for accuracy, because in 1947 it was clearly less than half of what Clarence turned over before the construction programme. This simply cannot be a fault in the design or construction of the work, which completely transformed what had been among the most obsolete docks in the country into a thoroughly good modern system, and it was not a question of price, either. It has to be the result of structural decay in the coasting trade, and it is well to remember that while, overall, the coasting trade had an "Indian Summer" after the war, in Liverpool in 1955 we find that the "tendency is one of continuing decline."[41]

While the above may seem an unreasonable use of hindsight in judging the success or otherwise of a programme of work begun before the Great Depression, in fact it merely illustrates once more the exceptionally problematical nature of strategic decision-making in a port authority. Decisions have to be taken which will be good for at least twenty years, and they nearly always have to be taken in an atmosphere of intense arm-twisting, both for and against, by the customers of the port. If some construction programmes failed to anticipate changes in the shape and size of ships, predicting the future course of particular trades was even more difficult. But it was what port authorities had to do.

Among shipping men it had gradually become an article of faith that coastwise shipping was a good thing, the best and most efficient way of forwarding imports and collecting exports for ocean-going shipping. In theory this was probably correct in that overside discharge to a coaster was quicker and easier than discharge to railway wagons or motor lorries, particularly when it enabled working the ship from both sides. In practice, sadly, the historical consensus that the coastwise trade never fully recovered after the Great War

---

[41]*Report of the Liverpool Steam Ship Owners Association* (Liverpool, 1955).

seems justified - and much of the evidence the historians have used was available to the Board when they took the decision to undertake the Clarence scheme.[42]

Ultimately, it was the long-distance lorry rather than the railways which finished off the coastwise trade. As early as 1947, forty-six percent of imports landed left Liverpool docks by long-distance lorry and 48.4 percent of exports arrived that way for loading. The coastal shipping market share fluctuated in the region of three percent. John Armstrong is doubtless right in asserting that the motor lorry made little impact in long-distance traffic in the early 1920s and that what impact it had was largely in local delivery, where it proved a positively useful adjunct to coastal shipping. By 1928, however, it could be suggested that it posed a much more visible threat over long distances, as did the better steam lorries.[43] Maximum speeds of around forty miles per hour were commonplace among the lighter vehicles - and rising: maximum *usable* speeds were rising proportionately faster as the country engaged in a huge programme of road improvement, permitting the use of pneumatic tires on "heavy" goods vehicles. Maximum permitted sizes and weights increased as well. Members of the Board who read *Engineering* would find in that year's issues a number of significant technical advances, including the introduction of a five-ton lorry driven by a sixty-five horsepoer M.A.N. diesel engine and improvements in transmission, springing and braking. As early as 1920, the admittedly somewhat visionary Liverpool City Engineer, John Brodie, had not only forecast the future triumphs of the motor lorry but also given some remarkable figures for the increase in its use.[44] But above all, if the Board could somehow contrive to be unaware of the potential of the long-distance lorry, why did it contribute towards the cost of building the East Lancashire Road?

The re-design of the Central Docks for the use of the coastwise trade was a well-intentioned project which was well-engineered, delivered on schedule and comfortably within budget. It was not a failure in the sense that nobody wished to use it, but it did not pay its way and it is very unlikely it ever could have done.

---

[42]But note "*fully* recovered." J. Armstrong, "Climax and Climacteric: the British Coastal trade 1870-1930," *in* D.J. Starkey and A.G. Jamieson (eds.), *Exploiting the Sea: Aspects of Britain's Maritime Economy since 1870* (Exeter, 1998), argues that recovery after both the wars was stronger than usually realised and that terminal decline came later.

[43]Sentinel produced a six-wheeler in 1929 and an eight-wheeler in 1930, both capable when fully laden of exceeding the then-speed limit of twelve miles per hour by a very large margin.

[44]On the principal roads leading out to Lancashire, some 3,500,000 tons were carried each year. J.A. Brodie, "Presidential Address," *Minutes of the Proceedings of the Institution of Civil Engineers*, CCXI (1920-1921), 6-29. By 1929 there were 296,000 commercial motor vehicles on the road. *Engineering,* 8 November 1929, 594.

This would have been acceptable had it provided a first-class adjunct to the ocean-going vessels, but this appears to have been just about the least successful aspect of coasting in and out of Liverpool. Even in the two immediate postwar booms, when there was terrible congestion in the port brought about by delays in removing imports from the quays, the coasters do not seem to have been the solution. It may be, of course, that had the Board kept its original promise promptly, perhaps by building the proposed floating stage at Waterloo, the pre-1914 growth in the trade might have been sustained. This would have left it in a much stronger competitive position, suggesting that perhaps the Board's earlier policy (shared by most other large ports) of always subordinating the interest of the coasters to those of the large vessels was, as alleged, so harmful that when the policy was reversed it was already much too late.

Chapter 4
Engineering, Management and Decision-making

The previous chapter raises a major question. How could the engineering of the Clarence modernisation project go so well, when the modernisation of Alfred Entrances on the Birkenhead side just a short while before had seen a breakdown in organisation which verged on the scandalous and resulted in fifty-percent overspending on each of the two major elements of the job, together with long schedule over-runs? The Works Committee appointed a Special Committee to investigate, which found terrible lapses of communication between the men in the office and the men on the ground.[1] It also found attempts to conceal overspending. It seems likely there was much else that it did not find, or at least record, including the following little episode.

When the Traffic Committee was engaged in a budgeting process in 1928, the following items appeared in its estimates for 1928-1929: Mersey Cattle Wharf, re-glaze slaughterhouse, cooling and chill room avenues; re-glaze Nos. 1, 2 and 3 sheds, Wallasey Dock and Glass House [lairage]; and re-glaze various panes in other adjacent sheds. Albert Foulkes, Chief Traffic Manager, suggested that as this work was "consequent upon the excavation and blasting in connection with Alfred Dock Works" it should be charged to those works. The same minutes record that "the Engineer-in-Chief is of opinion that only a portion of the damage was caused through blasting."[2] The list includes every building within 300 yards of the south wall of Alfred's south entrance, and the incompetence in the use of explosives necessary to achieve such destruction clearly implies serious failures in training, supervision and communication. Quite simply, the laying and detonation of charges had been entrusted to someone who was a danger to himself and to everyone else for a considerable radius.[3] There was an established

---

[1]The fruits of its meetings are in National Museums Liverpool (NML), Maritime Archives and Library (MAL), Mersey Docks and Harbour Board (MD&HB), Unbound Worked-up Papers (UWUP) A47.

[2]*Ibid.*, Worked-up Papers (WUP) 202/1, Traffic Committee, minutes, 15 April 1928. Newell's reaction sounds rather like the schoolboy plea in mitigation "Please Sir, it was cracked already." Since the cost does not show up later in the Traffic Committee's accounts, then either the Board agreed with Foulkes or it was "hidden" with some skill.

[3]Nor did the danger arise without warning: on 24 July 1901 Michael Letties, a labourer, blew himself to bits with five pounds of Tonite through "the wrong cable being attached to the battery." *Ibid.*, Staff and Wages (SW) 3/1.

professional literature on the use of explosives, which was useless if the person who determined the size, location and number of charges had not read it, including a paper specifically dealing with very similar work to that being undertaken at Alfred which described practices which, if carried out would have ensured a safer and cheaper outcome.[4] In the other major industries where blasting was routine, such as mining and quarrying, it was illegal to entrust untrained men with high explosives, but in engineering construction or demolition it was perfectly legal, if extremely foolish, to do so.

We may begin by dismissing the possibility that the difference lay in a failing in the Birkenhead Dock Yard, because although the staff there were in some senses partially autonomous, major construction projects were centrally managed and in any case, the Bidston Dock project on the Birkenhead side which proceeded almost exactly in parallel with Clarence was delivered on time and within budget despite difficulties which were not of the engineering team's making. For the same reason we may discount the excuse proffered at Alfred, of the low productivity of temporary labour employed under the Unemployment Grants Committee (UGC) scheme as compared with regular men, for UGC labour was common to all three projects. There is no direct evidence either way, but it is extremely likely that, as the excavation and walling operations at Gladstone neared completion and the labour force began to dwindle, some of those experienced men turned up again at Alfred anyway.

Some of the key personalities changed. Richard Holt succeeded Thomas Rome as Chairman of the Board and H.B. Gordon Warren filled Holt's former position as Chairman of the Works Committee. The minutes of the Works Committee were by this time extremely formulaic, and it is difficult to gain a flavour of the meetings, but no major change is immediately obvious. The first three items on the agenda were always inspection and approval of the Engineer's Order and Petty Cash Disbursement Books, permission to order goods or services and permission to sell off old materials, and it did not get much more exciting later, since none of the discussions (if any took place) were minuted. Such items as sale of old materials are in fact usually even less informative than might be expected, since the quantity and price are frequently not minuted and we have to resort to the committee support papers, where available. It is only when an item from the Works Committee was reported in the press or discussed at the Board and included in the *verbatim* minutes sequence (Discussions at Board) that we can identify individuals or appraise their styles.

In February 1928 Thomas Newell, Engineer-in-Chief, celebrated his sixty-fifth birthday; he retired on 2 July. Among those who wished to hold him entirely responsible for the troubles at Alfred, the most prominent was Hugh

---

[4]J.F. Ramsbotham, "Demolition of Old Masonry and Concrete," *Transactions of the Liverpool Engineering Society*, XXIX (1908-1909), 22-56.

Rathbone, scion of a famous old Liverpool merchant family. Rathbone took the then-orthodox view that whatever errors might be made by underlings, the Engineer-in-Chief was paid a very large salary to accept responsibility for everything that happened in his department. Rather in the manner that Harold Littledale had "gone after" Lyster Senior in the 1870s and 1880s, or John Hughes had occasionally castigated Lyster Junior, so now Rathbone repeatedly raised the question of Newell's competence, including questioning his being offered the customary (and one suspects rather nominal) post as Consulting Engineer at a retainer of £750 per annum, which he clearly regarded as a sinecure. As it happens, he was proved right, albeit for the wrong reasons, because Newell was in poor health and died before any further use could be made of his considerable experience and knowledge of the Dock Estate.

Like Jesse Hartley and G.F. Lyster before him, Newell had found a place in the Board's service for his son, who started as a draughtsman on 20 February 1914 and by 1927 had become Resident Engineer at the Gladstone entrance.[5] On 16 September 1927 the Works Committee notified him, along with two others, that with the completion of the Gladstone entrance there would be no further work for them after 31 January 1928. The other two (G.E.C. Callow and H.M. Horrocks) resigned with effect from 12 December 1927, but Newell clung on. This also clearly annoyed some members of the Board, and on 11 July 1929 he was reminded that he should seek to further his career "outside the Board's service." Nevertheless, the "hawks" did not have the necessary majority to sack him, and on 15 June 1933 he was finally given a permanent post as Superintendent of the Coburg Dock Yard. This was a slightly unusual appointment in that he was an Associate Member of both the Institution of Civil Engineers and the Institution of Mechanical Engineers, and superintendent posts had tended in the past to be held by "ex-rankers" - men who began their careers as apprentices rather than pupils.

If either the Works Committee or the Board was delighted to see the back of Newell Senior, they certainly concealed their feelings thoroughly; he was kept on full salary until 1 January 1929, and superannuated thereafter at £1150 per annum with, in addition, the controversial retainer as consultant engineer of £750 per year.[6] The clinching piece of evidence for their esteem of him is not so much the fairly conventional expressions of regret at his retirement or the even more conventional vote of condolence when he died, as in the choice of his

---

[5]A.G. Lyster had no son, but his nephew Osbaldiston Lyster worked in the Department for a short time. NML, MAL, MD&HB, SW/1/21

[6]*Ibid.*, Works Committee, minutes, 30 December 1927.

successor, appointed to take up duty on 2 July 1929.[7] From a large number of applicants, the Board chose Thomas Norfolk, formerly Principal Assistant, and as such presumably the least likely man in the entire profession to get the job if what the members wanted was a new broom.[8] Furthermore, Norfolk's former post was also filled by the internal promotion of Leopold Leighton, who would in his turn go on to be Chief.

There is no doubt that under Newell some of the work of the Engineer's Department was badly done. On 8 February 1929 the Works Committee heard that a recent inventory had revealed that the Engineer's stores showed a deficiency of stock against "the books" of £11,898 – and also contained materials to the value of £12,901 which did not appear in the books.[9] The procedures for recording the reception and issue of stores were labyrinthine and extremely time-consuming, and they quite clearly did not work.[10] The double discrepancy was said to have accumulated over the period of seven years, despite the fact that stock checks were supposed to take place every three years, and it had probably arisen because the quantities of materials issued were often estimated rather than exactly counted, measured or weighed. There is no suggestion of any dishonesty, just the Board's old problem of having administrative systems so complex that they had to be circumvented in order for anyone to get the necessary work done. The solution to the problem of the double discrepancy was to accept as correct the inventory of what was actually in the stores and alter the books to suit. Whether this actually meant *all* the books is unclear, but if it did then it would represent a week or so's work for several clerks to re-calculate the figures in all the various "analysations," as they were called, which stemmed from the actual receipt and issue of stores.[11]

---

[7]Only *fairly* conventional because the Chairman, in formally proposing Newell's superannuation, said "Of course we all know Mr Newell and we know his shortcomings and we all like him very much." Unfortunately he did not elucidate. *Ibid.*, Discussions at Board, 5 January 1928.

[8]*Ibid.*, Works, Committee, minutes, 18 May 1928.

[9]A similar situation also arose in the Marine Surveyor's Department.

[10]They had been marginally simplified in 1913: for details of the earlier system see A. Jarvis, "Attempts at Reducing the Expenditure of the Port of Liverpool 1836-1913," in P.C. van Royen, *et al.* (eds.), *Frutta di Mare: Evolution and Revolution in the Maritime World in the !9th and 20th Centuries* (Amsterdam, 1998).

[11]"Analysations" were secondary records which recorded the use of labour and materials, re-sorted according to various criteria.

Yet when the engineers actually got to do some real engineering, they did it well. The scientific study which they made, in collaboration with the Universities of Liverpool and Manchester, of the Mersey Bar and channels was really "sharp end stuff" that along with the fruits of some external consultancy would prove to be the beginning of a comparatively cheap and simple way of controlling the wayward sands of the port's approaches. The importance of this work can perhaps be judged from the estimated coal consumption of the sand-pump dredgers, which ran at 2750 tons per *month*.[12] The design and execution of the Clarence and Bidston projects were, as we have seen, good enough to see out the life of the trades for which they were intended.

A.G. Lyster had been early in the business of electric lighting and had installed some quite early electric cranes. Under Newell there was a prolonged campaign to electrify mechanical plant, such as graving dock pumps and quayside cranes. Within the financial limitations imposed by the need to complete Gladstone and the generally weak state of trade, it was successful. If, as the Chairman suggested, the accident rate of the workers should be one of the elements of an appraisal of a major engineering work, then Newell's performance at Gladstone was good. In 1924 (chosen as a "busy" year at Gladstone), for example, the total number of accidents in his department was 477, of which only 164 occurred at Gladstone. Three of the 477 were fatal accidents, none of which occurred at Gladstone.[13] The vast majority of the accidents there were "slight," arising from such traditional causes as treading on a brick while carrying a plank, resulting in a sore ankle and bruised ribs. If Newell is to be criticised on his safety record then it should be on the grounds of the considerable number of minor mishaps which occurred in the workshops among the practitioners of what should have been inherently much safer jobs than heavy construction work.

There is, of course, another aspect to successful dock engineering, the crucial one of doing the simple things well, which was one the marks of the

---

[12]This is from a tender reported to the Works Committee, minutes, 16 September 1927. When *Leviathan*, the largest of them, was laid up as an economy measure in 1931 the estimated annual saving was £40,447.

[13]It would be nice to give a percentage rate for accidents, but determining the numbers of men working for port authorities is notoriously difficult; see, for example, the controversy between Wardley and Jeremy and Farnie, whose rival estimates of the number of employees of the Port of Manchester differ by a factor of four. P. Wardley, "The Emergence of Big Business: The Largest Corporate Employers of Labour in the United Kingdom, Germany and the United States c. 1907," *Business History*, XLI (1999), 88-116; D.J. Jeremy and D.A. Farnie, "The Ranking of Firms, the Counting of Employees and the Classification of Data: A Cautionary Note," *ibid.*, XLIII (2001), 105-118; and P. Wardley, "On the Ranking of Firms: A Response to Jeremy & Farnie," *ibid.*, XLIII, (2001), 119-134.

greatness of Jesse Hartley and his team. Under Newell, for example, the ferro-concrete framing for the new shed at East Princes was of superb quality, its reinforcing bars exhibiting no rust when the building was demolished after nearly seventy years' service in a distinctly hostile environment.[14] The system may have made inefficient use of resources, but it produced the goods. Thc quality of execution at Gladstone, too, was excellent, resulting in the dock being still highly serviceable today.

In the case of smaller works such as the East Princes Shed it may be argued that a cheaper and less durable structure would have sufficed, but few dock engineers at the time, or their employers, would choose to store up maintenance problems for the future by building on a shoestring unless the financial situation left them no alternative. From at least the time of Jesse Hartley onwards, the preferred approach was to build things in such a way that they would last decades before needing anything more than a coat of paint.[15] It seemed, in the main, a wise policy, allowing future resources to be concentrated on meeting future needs yet to be identified. Another important benefit was that it minimised the loss of berthing space (and hence revenue) to repair works.[16] A persuasive case, perhaps, but one which failed to take account of the fact that by the 1920s "future needs" would not necessarily involve the continuum of growth which had characterised the port's trade for over a century. In particular, it failed to anticipate the continued decline of the coastwise trade, whose earlier expansion had always provided a use for, and a small revenue from, older facilities. That, however, would have required a quite remarkable degree of foresight, which one could not reasonably expect in an engineer and which was probably rather a tall

---

[14]I was there, watching it come down. It is not generally known among laymen that "concrete sickness" can be entirely prevented by careful and generous specification of materials – followed by hawk-eyed supervision to ensure that the specification is adhered to.

[15]Hartley's re-construction of the old semaphore telegraph system in 1841-1842 is a good example: the initial cost was high, but the previously high maintenance costs were vastly reduced for the rest of the life of the system. For details of expenditure, see NML, MAL, MD&HB, *Annual Accounts*, for a few years before and after 1841. On the telegraph generally, see F. Large, *Faster than the Wind: A History of and a Guide to the Liverpool to Holyhead Telegraph* (Liverpool, 1998).

[16]This was a serious cost to the Board, of which it was well aware but which it seems never to have quantified and which is difficult now to determine with any accuracy. I estimate that in a "busy" year approximately ten percent of the quayage of the port was in the hands of the Engineer – and that is not counting the mooring space for his floating plant. The issue is treated in greater detail in chapter five.

order for the Board as well – especially in a project which was specifically intended to help revive the coastwise trade.

If the Engineer's Department contained the requisite engineering skills, albeit hampered by an unspeakably bad office and accounting system, it seems reasonable to assume that the problems at Alfred could indeed be attributed to poor communication: that would certainly account for one critical failure, namely the absence of investigation of the site's history before the shovelling began. The estimates were drawn up by the Head Draughtsman, who had never been on site, and the Resident Engineer seems to have had hardly any contact with "head office." Just such a failure of communication seems to have been an underlying cause of the major accident at Vittoria Dock in 1909, when a large temporary dam collapsed, inundating the site and killing fourteen men.[17]

Hugh Rathbone (and Colonel Buckley and Henry Fernie and one or two other members) thought that the responsibility for the Alfred incident rested with the Engineer-in-Chief, but after the special sub-committee investigated the affair the Works Committee recommended that matters be let rest. Mr. Bibby, in discussion of the Report, said that:

> I agree that the Engineer has been grossly wrong but I will say this on their [sic] behalf – that they inherited a slack and loose system – which ought to be deprecated. In the past we had one engineer or another who has been the supreme authority and who did not look upon amateur opinion as worth very much, and the Draughtsman looked upon him as a little god...[18]

There was one small problem with Bibby's deprecation of the inherited system: born in 1846, he had been a member of the Board from 1892 to 1900 and then again from 1905 to the time he spoke. Richard Holt, Chairman of the Board, had first been elected in 1896 and had been a member of the Works Committee from 1902 to 1927 (chair between 1919 and 1927). The fact is that if these two gentlemen had made too much fuss about "inherited" problems they would have put themselves in a very exposed position, and the "hawks" were in little better case: Colonel Buckley had only joined the Board in 1928, but Henry Fernie had been there since 1910 and Hugh Rathbone since 1905. If the system was so bad, why had none of them lifted a finger to improve it during their decades of service? In the event, the Head Draughtsman, "who is no longer in the service

---

[17]For further details, see A. Jarvis, "Forgotten Heroes: The Dock Construction Workers," *Northern Seas Yearbook* (1997).

[18]NML, MAL, MD&HB, Discussions at Board, 25 February 1926. Presumably he means the singular form throughout, but this is not clear.

of the Board," and the Second Assistant to the Engineer, a youngish man named Gibson Smith, were held chiefly to blame, and the latter was passed over for promotion to Principal Assistant in succession to Thomas Norfolk when Leopold Leighton - in the Department's arcane terminology, only "An Assistant to the Engineer" - was appointed.[19]

From the evidence in the minutes, it seems that two quite simple reforms account for the fact that the problems at Alfred were not repeated in either the Clarence Modernisation Scheme or the Bidston Dock construction. The first was an entirely predictable move intended to give the Works Committee more knowledge of and control over what was actually happening on the ground. From 1836 onwards, Jesse Hartley had produced brief but informative monthly reports, but, so far as may be told, this practice lapsed under G.F. Lyster and was only resumed in the 1890s under A.G. Lyster. These monthly memorandum books (which are *not* copies of internal correspondence as modern use of the term might lead one to believe) were, however, quite sufficiently informative to start with and probably gave members more information than they could routinely assimilate.[20] It is difficult to see what difference the new format could have made.

The second reform is potentially much more telling. Richard Holt questioned whether the Board was not partly to blame for giving the Engineer's Department insufficient time to work up proposals in a sufficiently detailed manner. He suggested that schemes should not be authorised until a full set of working drawings, as distinct from "plans," were available for estimating purposes. It is stated above that "no immediately obvious change" was noticeable in the conventional forms of words used in the minutes of the Works Committee. There were, however, two changes which were, as stated, not immediately obvious but which become significant when viewed in their proper context. Clear distinctions came to be drawn between plans and working drawings and between estimates and final estimates. This had always been necessary when work was being undertaken by a contractor, but in-house work had often been undertaken in a much less formal manner, leaving substantial power - and discretion - in the

---

[19]*Ibid.*, 11 July 1929. The surviving personnel records cast no light on the Head Draughtsman's departure, but Gibson Smith was eventually forgiven: in 1941 Leopold Leighton was promoted Engineer-in-Chief and this time Gibson Smith was promoted from being "An Assistant" to being "The Assistant." (i.e., Principal Assistant after the job title had been changed). NML, MAL, MD&HB, SW/1/21.

[20]A. Jarvis, "The Members of the Mersey Docks & Harbour Board and Their Way of Doing Business," *International Journal of Maritime History*, VI (1994), 122-139, considers this issue in much greater detail for the earlier period; its chronological coverage is extended below.

hands of the top echelon of "blue collar" men.[21] These men, whose ancestry as crucial figures in the dock construction process stretches back a long way, were the ones who oversaw the actual construction, and with few exceptions they did it superbly.[22] It was probably the Head Draughtsman's defective interface with these men, rather than with the Resident Engineer, that caused the problem. They were doers rather than estimators, men of a different culture and usually of a different social class as well.

Seen from T.L. Norfolk's point of view, as the new Chief of a clearly troubled department it was essential for him to get those channels of communication functioning properly again. But his short-term goal was simpler: it was to avoid any more fifty-percent overspends, and if overspending is Scylla, then Charybdis is under-estimating. The metaphor breaks down, however. What he spent might not be under his control, for as he knew only too well it is possible to encounter expensive surprises, whether natural or man-made, underground. He could, however, control what he estimated, and as appears in chapter three, there is little doubt that at Bidston and Clarence his team simply wrote a list of everything that could possibly go wrong , included it all and then added another ten percent for contingencies on top: he moved Charbydis. How often may we see well-intentioned economy measures side-stepped, in this case by an engineer doing something not normally expected of his profession, namely deliberately and heavily *over*-estimating.[23]

Perhaps the members uttered loud sighs of relief when the Clarence and Bidston projects apparently vindicated their "reforms," but what had really happened, and as businessmen they should have seen happening, was that a crafty and still over-powerful Engineer had pulled the wool over their eyes. But we cannot entirely blame Norfolk for that: the Board's Standing Orders state perfectly clearly - emphatically even - that responsibility for the Estate and all works upon it, including "oversight" of such works, rests with the Works Committee. The situation in the 1920s begins to reflect that in the 1880s, when G.F. Lyster seemed able to circumvent any effective oversight of his activities by

---

[21]The use of the term is figurative, referring to their having a trade rather than professional training. In reality an ordinary foreman wore a white shirt and a "blocker," and there were Head Foremen and a Principal Foreman above him.

[22]The first one of whom much is known is John Millar, appointed Principal Foreman in 1846. Brief particulars of his career are in NML, MAL, MD&HB, SW/1/8. The noteworthy point is that he was paid as much as John Ellacott, Principal Assistant to the Engineer.

[23]Though, of course, A.G. Lyster's practice of leaving a bit of "fat" in the estimates in order to preserve his favourite parts of a scheme from cuts was fairly widespread. What was different here was the scale of the over-estimating.

the members. The same question therefore arises, which is how it came about that a committee of a body supposedly made up of twenty-eight of the best business brains in the city could fail in its duty to monitor and control the activities of one newly-appointed principal officer. In fairness to Norfolk, if he was going to mend the internal ills of the department, he desperately needed to improve its image with the Board members and beyond, and that he appears to have done.[24]

There was also the issue of indecision, of which the outstanding example is the construction of Gladstone Dock, described in chapter two. For present purposes the key points are simply that the Board could not decide what sort of ship the new dock was for or what its basic orientation to the river should be, and that between 27 July 1904 and 16 June 1905 six different designs were requested and produced. When work began to yet another design, the money ran out after little more than fencing the site had been completed, and when it re-started in 1911 it was to a design completely different from any previously considered. This was not the most effective way of employing the costly skills of a design office. In this case, contrary to the opinion of Mr. Bibby, at Gladstone and Clarence it was when the amateurs got out of the way and let the professionals get on with it that things began to work as they should.

This question has been investigated for the earlier period by the simple method of analysing the biographical files kept on members of the Board, and the same method is applied again here, looking at the 101st to 150th members elected, from William Adamson (elected 1895) to John Tilman (elected 1925).[25] The last of this cohort to leave the Board was Sir Robert Johnson in 1948, so that in the second group as in the first, one obvious reason for poor performance could well be that members were too old when elected, were staying in office too long, or both; in short that the Board was a conservative gerontocracy. There is certainly *prima facie* evidence for this: of the fifty members, the dates of birth of forty-two were on file and of these, twenty-seven were aged over fifty (nine were over sixty) when first elected; just four were under forty. Of the total, ten remained more than twenty-five years and twenty-two – more than half – remained beyond the age of seventy and three beyond eighty.

Of the full sample of fifty, twenty-five were only voted off by the Grim Reaper, including Sir Richard Holt, who gained the record by staying forty-five years. The impression is strengthened when we realise that of the twenty-five for whom there was life after the Board, every last one resigned: not a single

---

[24]Certainly the accounts of the Clarence and Bidston projects in the weekly-journal *Engineering*, which could be fairly acerbic at times, are very complimentary.

[25]These are known as the Antecedents Files, a sub-group within the UWUPs. In the interests of consistency, this source, and only this source, has been used throughout this section.

"elective member" was voted off, and contested elections were virtually unheard of. The rate of turnover was also, of course, much reduced. Between 1858 and 1894 a new man joined the Board, on average, every 4.3 months, while between 1895 and 1925 that rate fell to 7.2, a decrease in turnover of almost exactly sixty percent.

Table 1
The Age Profile of Members Elected to the Board between 1895 and 1925

Age at Election	Number of members (N = 42)
<30	1
30-39	3
40-49	11
50-59	18
>60	9

Table 2
Length of Service

Years on Board	Number of members (N = 50)
1-5	8
6-10	6
11-15	16
16-20	8
21-25	2
25-30	6
>30	4

Note: Two members died prematurely in service, Charles Paynter on the *Lusitania* and Charles Booth in a fall from a window.

Source: National Museums Liverpool (NML), Maritime Archives and Library (MAL), Mersey Docks and Harbour Board (MD&HB), Antecedents files, various years.

If gerontocracy was indeed a problem, then it was a bigger one than is immediately apparent. In the early years of the Board, Members were elected to the Standing Committees according to what their principal occupation might suggest they were interested in or good at. But it soon became clear that some committees were more important than others, and that the most important – because it oversaw the spending of far more of the Board's money than any other – was the Works Committee. Close behind, and sometimes jostling for control, came the Finance Committee. A recognised but unpublicised hierarchy of committees developed, with newly-elected members being appointed to less important committees (such as Traffic) and gradually "promoted" over the years, some finally achieving membership of one or even both of the "top" committees. What this meant was that if the Board as a whole was rendered less effective by the old age and/or "staleness" of the members, the damage would be concentrated in the two most important of the committees. And so it proved: only half of our sample of the Board at large stayed in office until they died. Of the fifty members sampled, twenty-six were eventually elected to the Works Committee and nineteen of those twenty-six - or fifty percent above the average for the whole of the Board - died in office, including William Clarke, at the age of eighty-seven.

Of course, it is true to say that many people can and do carry on doing useful work well past the age of seventy. It is also true that in some cases very long experience may be a more valuable quality than innovation or energy - perhaps even than intellect. We cannot assume that simply because the average age of members of the Board was high and rising that the Board was *necessarily* an ineffective body. But that is not where we started; we started from the suggestion that the Works Committee was, as a matter of fact, not adequately discharging its duties of supervision. It seems that the gerontocracy hypothesis may contain at least some truth.

It was hinted above that the electoral process may have become stultified, and it is now positively suggested that it had, in two ways. The first was that although the members of the Board as originally constituted were expected to act disinterestedly, we find an element of "representation" creeping in from an early stage. John Laird, one of the original nominee members, was effectively put there to represent the interests of Birkenhead, just as Hugh Mason was nominated by the Manchester Chamber of Commerce in 1872. By 1905, the Liverpool Corn Trade Association could openly remove their sitting member, Harvey Woodward, and replace him with another nominee. By the time our cohort of members was well settled in power it was automatic that if a representative of a particular trade died or resigned it was the trade association which would nominate his successor and that nomination would be unopposed.

Of course, it may be thought that this, for all its falling outside the spirit of the 1858 Act, was an efficient way to run things. It maintained a balance of power between different interests on the Board which reduced (though by no means eliminated) internal strife of a particularly unproductive nature. Unfortunately, it took no account of changes in trade. For example, petroleum products had been imported into Liverpool since 1862, with the Board's first bulk installation opened in 1891. It was easily the fastest-growing trade in the world, increasing nationwide by a factor of fifty between 1859 and 1889. Clearly, such a trade needed a representative on the Board, and it got one in the shape of the aptly-named William Smellie - elected in 1915! Now "representation" as a principle of election was either good for the Board's conduct of business or it was not; it was most unlikely to be neutral. If it was good, then someone from the oil trade should have been elected by about 1890 at the latest. But Smellie's election caused grave offence to the timber trade, which "lost a seat" to accommodate him. When Smellie retired in 1928, the timber trade had not forgotten old scores, but the seat now "belonged" to the general brokers, so the Chairman and the General Manager had some careful caballing to do; eventually they persuaded the general brokers to do the decent thing and nominate a timber man. Stuart Mountfield states that in 1923, "constitutional problems were in the air" and goes on to explain that "members had been shaken by the demands of the sugar trade to be represented,"[26] demands which went to such extremes that it was necessary actually to hold a contested election. Now whatever Mountfield's weaknesses as an "insider historian," he exhibited in full measure one of the strengths of insiders, namely a great "feel" for what was happening, and if he says that holding an election was a "constitutional problem" or that members had been "shaken" by such an occurrence, we can almost certainly believe him.

In short, it seems fairly clear that any question of electing someone on the basis of ability alone was long dead, but it is by no means clear that this constitutional evolution within the Board was beneficial in either reflecting the views of port users (which was its primary purpose) or of maintaining a peaceful state in which the Board could better deal with the pressing issues before it. It may be that it was simply a way of avoiding facing up to change.

At this point we must recall that not one of our fifty members was voted off the Board and very few indeed among the first hundred members were either.[27] What this meant was that it became effectively impossible to vote anybody at all off the Board, no matter how useless they might be, because there could be no weasel-worded explanation. Voting somebody off during the 1920s would have been an unambiguous statement that the departed had been not just a bit past his prime, or wanting to spend more time with his great-grandchildren, but *uniquely* useless. Who would do that to a kindly old man like Joseph Hodgson, who died in office at the age of seventy-seven?

Members continued to have a wide range of commitments which might interfere with their duties on the Board, but there was a noticeable shift from the situation among the first 100 members, where a sense of public duty was very evident. In the earlier sample, two-thirds were Justices of the Peace, while in the later less than one-third were. Less than one member in eight had active church commitments compared with exactly one-half in the earlier sample, and participation in the management of charities fell to seventeen members. Only six

---

[26]S. Mountfield, *Western Gateway: A History of the Mersey Docks and Harbour Board* (Liverpool, 1965), 142.

[27]Including Harvey Woodward. The Corn Trade Association simply declined to nominate him for re-election because, most unusually, it insisted that its representative must be actively engaged in the trade, and Woodward had recently retired.

out of fifty were active in local or national politics. The sense of needing to be prominent in "public life" clearly dwindled considerably, and we find instead that more members are or were involved in sports of one kind or another. But above all their interest in trade associations grew. Of the first 100 members this information was available for fifty-two, of whom twenty-five were involved in trade associations. Those in the later sample of fifty aggregated seventy-three memberships, representing probably both a cause and an effect of the increased influence of the associations on the Board.

This influence seems to have become more extensive than the remarks above can convey. In 1921 Newell participated in a discussion in which one of the issues was how dock engineers were to know how large new facilities needed to be, a question which had occupied their minds for decades.[28] It was suggested that rather than asking shipowners aboout the type and size of vessel they might be building in a certain number of years' time, they should ask the shipbuilders. That was clearly untenable, since shipbuilders built what shipowners ordered, but Newell went a stage further: "He thought the direction in which they had to look for the future was to the trader who was going to utilise the merchant service." The consultations which preceded new construction had long included trade associations, but to suggest that they were or should be the principal channel of information was new and was probably connected with the shift of influence within and without the Board suggested above. The reason for the inclusion of the cautious "probably" is that in the case of fast passenger liners Newell also thought it necessary to "backtrack" in the same way by looking to the market for the services they offered. The radical suggestion that port authorities should independently verify the market research of the shipowners might have been sensible, but it rather muddied the main issue.

One "charge" of which we can exculpate members 101-150, as well as their predecessors, is that of succumbing to the "clogs to clogs" syndrome, in which grandfather makes a lot of money which his son wastes in trying to become a gentleman, and grandson, probably hindered by an expensive "anti-industrial" education in the classics, becomes a complete layabout. Only one of the fifty (Edward Rayner) was "a sportsman" - and his will was proved at over £118,000 gross, suggesting he was not unsuccessful. Only one member lived out of town, and he was the "Manchester Representative." While all the members were

---

[28]In a discussion of H.J. Deane, "The Development of Cranage Facilities for Discharging Vessels of the Largest Size," *Proceedings of the Institution of Civil Engineers Conference* (1921), section V, 21-24. On ship sizes more generally, see chapters two and five.

referred to as gentlemen, I have not found a single instance of a member giving "Gentleman" as his occupation.[29]

It would seem, however, that a combination of a cosy environment, a high average age and a low turnover of members, coupled with still considerable outside commitments of time and (declining) energy would be quite adequate to explain poor policy-making and probably worse performance monitoring. It may be that what is required is not an explanation of how some things went badly but rather how some other things, under such handicaps, went well.

One distinct possibility is that the performance of the officers was more important than the members would readily admit. A.G. Lyster, for example, had a far-from-flawless record, but when the urgency of the Gladstone Graving Dock project forced members to leave him alone to get on with it, the result was a fine job delivered within schedule and within an estimate which Lyster had himself voluntarily reduced. One begins to wonder how a price could be put on all that Board indecision beforehand: one thing is certain, which is that it would only begin with the price of the wages and salaries of the people producing endless different designs; there would also be a heavy effect on departmental morale, because engineers like making and building things.

A somewhat similar situation applied with both the Clarence Modernisation and the Bidston Dock schemes. Here the urgency was directed to meeting the schedules of the UGC, and somebody, probably Sir Lionel Warner, clearly suspected that this display of government munificence would not last indefinitely. So it proved: the Committee, having been established by statute for a three-year term, was simply allowed to expire in 1932. This explains the willingness of the members to let Warner get on with making the arrangements to get work started rapidly, and then - despite their intentions after the Alfred affair - to leave Norfolk to get on with the engineering.

There is, however, no evidence that members interfered unduly at Alfred, which pushes us again towards the view that Hugh Rathbone may have been right, and that Thomas Newell may well have been at least partly responsible for the systems failure of his department. If he was, then he could offer two very sound excuses. The first was that the 1911-1913 Special Committee had presented its final report shortly before he was appointed and had concerned itself primarily with pay, conditions and the "need for work;" such sorties as it made into man-management issues were concerned with long-standing problems like the floating plant. He could, therefore, reasonably assume that the system he inherited was satisfactory, at least as far as it applied to major building and rebuilding projects.

---

[29]When cited specifically as an occupation, this is a euphemism for "rich layabout."

The second is that while the Alfred project was large, it was insignificant compared with the completion of the Gladstone system. Gladstone was no recipe for peace of mind; as originally designed it bore an estimated cost of £3,000,000, but the changes described in chapter two meant that a thumping overspend was inescapable before one takes into account the effects of the Great War. The eventual cost was approximately £8,000,000, and much of the time it was difficult to plan far ahead because of the unsteady state of trade and hence of dues income. Between 1921 and 1929 the proportion of the Board's dues income which disappeared in interest payments rose by almost a half, from just over forty percent to just over sixty percent. Yet everybody knew that slowing or ceasing work, leaving a hole in the ground which had cost millions already and would not earn a penny until complete or virtually so, was not saving money but making matters worse.

Gladstone was incomparably Newell's greatest problem, and he would have been less than human if he had not given it the lion's share of his own personal efforts and sought to deploy his best assistants (among whom he presumably numbered his son), foremen and workmen there. One bright young man who worked as an Assistant Resident there was J.D.J. Saner, who left to work for ICI when Gladstone was finished, returning in 1930 to be Resident for Bidston Dock. In 1958 he was appointed Engineer-in-Chief.

Even if proven, however, this suggestion that Gladstone robbed Alfred of the most competent people is not in conflict with that above to the effect that the main problem at Alfred was one of system and communication. The men who were – had to be – at the centre of the Chief's thoughts and attention were those at Gladstone, which would naturally attenuate communication throughout the rest of the department. By the time he retired, Newell was a tired man, aged beyond his years, and it seems likely that it was Gladstone rather than Alfred that made him so.[30] Be that as it may, he was demonstrably incapable of maintaining overall control of both projects at once.

This was a situation which the Works Committee should have been easily capable of noticing and which it could equally easily have corrected by placing a senior engineer (presumably Norfolk) in charge of coordinating the Alfred project. This had been standard practice on many earlier projects, and Fielden Sutcliffe seems to have spent most of his career (1873-1909) in just such a coordinating role in various parts of the Dock Estate. To the question "could a new Fielden Sutcliffe have saved the Alfred project?" one can only answer "probably." While by no means all the projects he worked on were entirely successful, it seems that it was his role which became the missing link in the department in the 1920s. It also seems that a less settled and more alert Works

---

[30]Ironically, John Hartley had gone the same way in 1861, and one of his leading headaches had been the bad ground in the vicinity of Alfred Entrance.

Committee might have seen the warning signs and insisted on someone fulfilling that role instead of leaving Newell to it. Yet the committee chairman at the time was Richard Holt, who appears by all normal criteria to have been a man of great ability backed up with a high degree of conscientiousness, so in a measure the mystery remains, and that is the fact which truly establishes where the guilt lay.[31] The question should not have been left for historians to chew over: it should have been resolved by the members at the time, in order that never again would organisational failings in the department result in problems like those at Alfred. It was not and they did.

A Putative General Explanation

It was not just in engineering matters that the Board took decisions which look strange or stupid to us. A small but classic example is the imprisonment of a fisherman for persistent trespass, as related in chapter eight. The man was clearly doing no harm to the Board's property and the initial legal action, *a fortiori* the seeking of a high court injunction, seems a course engendered not just by caution or meanness but by folly. Members' reasoning in attempting to exchange the site of the obsolete Chester Basin for the right to block up the public right of way at Gower Street (which runs between Albert and Dukes docks) is not immediately obvious either: they knew from the land sales at Georges Dock in 1900 that a waterfront site had considerable and lucrative development potential, whereas Gower Street had no evident value except in consolidating their existing control over the land surrounding two small and obsolescent docks.[32]

The Board continued to make objections to major schemes proposed by other bodies, including (for a time) the Mersey Road Tunnel, whose potential benefit to the operation of the docks was enormous. In 1921, it had objected to many clauses of the Liverpool Corporation Bill, its petition against running to twelve pages and getting down to such matters of detail as the fastening of any tramway or other electrical cables which might be attached to Board property under the terms in the bill. It also opposed public measures, including the Criminal Law Amendment (No. 2) Bill of 1920. This contained a clause forbidding the use of docked vessels as brothels and required port authorities to take steps to prevent "persons resorting to any port, dock or harbour for purposes of prostitution." Following lobbying by the Board and the Dock and Harbour Authorities Association, the government, anxious for an easy passage of what it considered more important parts of the bill, withdrew the clause. This was

[31]Though some readers might think his attitude to advertising, mentioned in chapter eight, was rather old-fashioned.

[32]NML, MAL, MD&HB, Parliamentary Committee, minutes, 19 January 1923.

reported to the Board's Parliamentary Committee on 4 April 1921 and elicited a most uncharacteristically acerbic minute to the effect that there would now be "no obligation on the part of Dock and Harbour authorities to interest themselves in the morals of those who frequent the docks."[33]

Similar attitudes seem to underlie the wordy contest the Board fought with Lever Brothers between 1920 and 1923. Lever, as mentioned in the context of barge traffic in chapter seven, received considerable amounts of raw materials by barge at its "dock" (in reality a small river quay on the banks of a widened-out Bromborough Pool) at Port Sunlight, involving double and sometimes triple handling of vegetable oils, palm kernels and the like in barrels. What it wanted was to be able to bring fairly large ships direct to its works, with particular reference to bulk handling of both oils and seeds. It might be expected that the Board would take a dim view of this, because it would either involve a major customer (Elder Dempster) transferring some of its trade from the Board's docks or, more likely, Lever operating its own fleet with at least equally damaging effect. Yet when the Lever Brothers (Wharves and Railway) Bill, 1920, went before Parliament its effect was astonishing: the Board did not only collude with its old enemy, the London and North Western Railway Co., and its subsidiary, the Shropshire Union Railway and Canal Company, it did the unthinkable and worked closely with its most bitter foe, the Manchester Ship Canal Company, even to the extent of having the same barrister lead for both their petitions against the bill.[34]

What Lever Brothers was proposing was a river wall with a T-jetty extending to the deep water of the Eastham Channel, and most of the evidence given concerned the likely effects of this on the channel. The contention was that by a series of knock-on effects this would risk damage not just to the Eastham Channel but to the whole river system. Much was made of the so-called "Lyster Line," a notional boundary beyond which no construction into the river should be allowed, which had been posited by A.G. Lyster in the context of the reclamation of part of Tranmere Bay for the construction of the new Laird shipyard, and it was claimed that Shodwell Point, where the jetty was located, was exactly the most harmful possible spot. The Board also made much of the fact that Lever Brothers, as an entirely commercial concern acting for its own profit, was providing no facilities or benefits for the public, and that if any such facilities were necessary (which they were not) the Board should provide them.

---

[33]It should perhaps be explained that while the discussions in committees could get quite heated, and are so rendered in NML, MAL, MD&HB, Discussions at Board, the formal minutes are normally extremely deadpan.

[34]There is extensive correspondence on the subject in NML, MAL, MD&HB, Legal B61.

These arguments were particularly weak. The private nature of the new "port" was in fact the result of previous negotiations with the Board, which had persistently, if understandably, worked to maintain, so far as it could, a monopoly of public facilities on the river. Given the concession that it should be a private facility, it chose to represent it as profiteering. That the facilities *were* required was self-evident, as was the fact that the Board was desperate to complete Gladstone and would certainly not be investing at Bromborough for a while. It had also turned down the opportunity to buy the site of the proposed development, which had been offered in 1905. The conservancy arguments, like all such claims, at least until the construction of the Second Mersey Model (see chapter six), were totally inconclusive, and advice based on the Board's in-house expertise was open to question in light of the continuing and conspicuous failure to deal with Pluckington Bank.[35] On 2 July the Parliamentary Committee heard that the Acting Conservator had withdrawn his opposition to the scheme. The Board soldiered on, lost, and the Bill received the Royal Assent later the same year.

What happened next seems incomprehensible. One of the counter-proposals made by the Board and the Ship Canal Company, supposedly to minimise the conservancy damage, was that instead of a jetty, an impounded dock should be built in the mouth of the Pool. This was not possible owing to the opposition of Price, of patent candle fame, which owned a crucial part of any feasible site for such a dock. But things now moved quickly, and Lever Brothers, having purchased a controlling interest in Price, decided that an impounded dock would be a better idea anyway. Accordingly, it went back to Parliament to gain sanction to abandon the previous scheme and to build a dock instead. Despite its previous loss, and despite the fact that a dock had been its preferred suggestion, the "usual suspects" turned out again with their petitions. Thomas Newell was the Board's only witness and, closely questioned about the problems at Pluckington Bank, was forced to lie in his teeth by claiming that the situation was improving and in sight of solution when it was demonstrably neither. The experience of the Board's "lead," Mr. F.G. Thomas, at the close of proceedings was a barrister's nightmare, as he faced a skilled opponent while putting a threadbare case. He got shredded. There was no question whatever of the objectors succeeding, and so much is evident from reading the opening speech of Lever's Counsel and the increasingly irritable tenor of remarks made by members of the Committee. It is not really necessary to read the evidence: the whole affair was quite clearly a waste of time and money.

---

[35]It is, of course, possible to check the outcome by use of hindsight, in the shape of the annual *Report of the Acting Conservator* for the years shortly after the dock opened, and these show the channels varying slightly from year to year in breadth, width and depth – just as they had done before.

Yet, as we have seen, the members of the Board were not stupid, even if they sometimes did stupid things. Why on earth did they do things like get a harmless fisherman gaoled or fight unwinnable enquiries, wasting money and making themselves highly unpopular? The answer may be given in one word: trust.

The Board had a problem right from the start with the "corporate identity" of the port authority. Because its statutory name ended in the word "Board," it was sometimes difficult to differentiate between the members when acting collectively and the authority itself. This led to the use of the expression "This Trust" when referring to the latter. What is in a name? It is here suggested that if they called themselves "This Trust" long enough they would come increasingly to think of themselves as a trust in terms other than the purely technical, that is to say as a body which held an inheritance which it was morally required to hand on to its heirs certainly in no worse a condition and preferably in a better one than that in which they received it. This was what gave such particular point to a remark of Mr. (later Sir Alfred) Read in 1905 to the effect that "it is not that you have neglected the coasting trade but that you have neglected your own estate in these central docks." Read knew how to hit where it hurt.

Seen in that context, some of the curious behaviour mentioned above becomes much more comprehensible. The incarceration of the unfortunate fisherman resulted from the Board's concern that if people were allowed unrestricted use of their piece of potentially useful development land, gradually by custom it would become a public right of way. The Mersey Docks & Harbour Company had actually had exactly this problem when laying out the new ro-ro berths at what is now known as Twelve Quays: Tower Road had been a *de facto* right of way for so long that the company faced stern opposition to returning it to operational use. The purchase of the land at Tranmere had itself been a piece of protection of the Board's potential long-term interests, for it was basically a pre-emptive purchase to ensure that no further attempts at competition from river quays or jetties there would worry its successors in the way that it had been worried. In other circumstances the Board was prepared to be quite generous over parts of its land for which it had no immediate use, allowing it to be rented for things like sports grounds at modest price – but always for short terms.[36]

The attempt to close Gower Street may be seen in a similar light. In 1900 the Board had purchased the almost totally obsolete Dukes Dock property from the Port of Manchester: just about its only use was the storage of barge-borne grain in its venerable warehouses. But it was an enclave of non-Board property

---

[36]See A. Jarvis, "Land Policies in the Port of Liverpool, 1857-1930," *International Journal of Maritime History*, XIV, No. 1 (June 2002), 115-133, for further details.

within what the Board wanted to be a simple boundary: everything between the Dock Road and the river should belong to it. It made everything simpler, and that simplicity was itself an asset. Gower Street was an anomaly: a city street that passed through the Dock Estate from one side to the other. It was also the access to the statutory public promenade to the river side of Albert; stop up the access and the promenade ceases to be a problem too. There was probably no policy of the Board which made - and still makes its successor - so unpopular as its attempts to prevent ordinary people walking along the bank of the river watching the ships. But the Board saw no need or wish for love: what it wanted was respect and stability.

And so to the great absurdity of Bromborough. The members in 1920 had inherited a port which dominated the estuary and whose immediate forebears had purchased a good deal of land, some of it at quite high prices, at least in part to ensure that competitive ventures like Garston would not be repeated. They had participated enthusiastically in the Tranmere Bay project for Laird because it tied up an awkward stretch of foreshore which had already attracted several potential competitors; with a shipyard built on it, their successors would not have to face that problem again. More topically, they had spent money on the Mersey and its channels not just sufficient to maintain it as they found it but to leave it as readily navigable as it had been before but by much larger ships, for the benefit, as they saw it, of the whole nation. As emerges in chapter six, this was a huge undertaking, but it was a matter of duty. Is it taking matters too far to suggest that, in that frame of mind the defence of the arbitrary and unproven "Lyster Line" could take on sufficient importance to justify to members the really rather ridiculous objections to the 1923 Bromborough Dock Bill? If so, then the fact that Lyster had been severely criticised by the Works Committee for his miscalculations in the dredging of the bar and his misplaced optimism in the matter of keeping Brunswick Entrance passable despite the Pluckington Bank would be rapidly and completely forgotten.

We could go on: the idea of trust encourages a willingness to spend on durability, as remarked on in the context of the Princes Dock sheds mentioned above. It justifies the time and money spent on careful scrutiny of bills, public and private, and the frequent petitioning of Parliament against anything which might be held to damage, even potentially, the interests of the trust. That might be a fear that the Birkenhead Docks branch of the Mersey Tunnel was to pass under Morpeth Dock at an insufficient depth, risking possible structural damage, or a realisation that the Board could be encumbered with a liability to employ extra police to clamp down on prostitution on the docks.[37] In 1927 the Board

[37]There is some irony here in that it did not object to its depth beneath the river wall at Pierhead, and a section of that failed owing to subsidence during excavation, resulting in lengthy litigation. NML, MAL, MD&HB, Legal E11, E12 and E13.

objected to no less than three local authority bills simultaneously, in each case for fear of sewer outfalls increasing the its dredging costs, as discussed in chapter six. Naturally, such objections did nothing to make it popular with local authorities, some of which were not above objecting to Board bills in order to extract some quite unconnected right or benefit in exchange for withdrawing their objection.[38]

It is important to realise that this tentative explanation relates to aspirations, not performance, so that its validity is not necessarily jeopardised by failures such as the passage of the Bromborough Dock Act. As the earlier part of this chapter shows, the Board won some and lost some, and things had long been so. In the case of the failure to control Pluckington Bank, for example, it had tried its best, was still trying, and had no grounds, on its own terms, for self-reproach. It was furthering the interests of "The Trust."

Let me conclude with a trifling example where the notion of trust seems to explain another bit of curious behaviour. Some members of the Board were strongly opposed to the vulgarity of renting space on port structures for advertising posters, which also attracted outside criticism.[39] Yet when the Advertising Regulation Bill was deposited in 1924, proposing to give local authorities control over poster sites, the Board lobbied for the exemption granted to railway companies for posters which could only be seen from inside stations or other premises to be extended to port authorities, and on 26 November 1924 the Parliamentary Committee heard it had been successful. The advertisements with which the Board was concerned were already limited to those supplying simple information, were located at the point of provision of the service advertised, and total income from this source in the year 1928-1929 amounted to £63 9s 8d. Two elements of the trust ethos emerge here: to reserve rights to itself and its successors, regardless of monetary value, and to resist any increase in the powers of local government *vis-a-vis* port authorities in general and itself in particular. It now seems a strange way of looking at the business world, but it was at least consistent, if not wise, at the time.

---

[38]For example, Wallasey Corporation objected to the Board's 1923 bill in order to gain direct representation on the Board – something the Board had long resisted. The Board made no concession: Wallasey petitioned and was ruled to have no *locus standi*. *Ibid.*, WUP 60/16. In fairness to Wallasey one should add that this was a time-honoured tactic which plagued (for example) engineers during the "canal mania" of the early 1790s.

[39]See chapter eight.

Chapter 5
Making the Customers Happy

The nearest the Mersey Docks Harbour Board had to a "Mission Statement" was the requirement in the Mersey Docks and Harbour Board Act 1857, which established it, that it be a non-profit trust governed by Board members elected by the users of the port to work in the interests of the users. While the Board was more successful in some of its endeavours than others, being popular with the customers was among its weakest suits. It was perceived, often unjustly, as a body dedicated to its own interests, unreceptive to the suggestions or complaints of others and given to "stashing" funds which, in the customers' view, should be applied to the reduction of dues. This perception was partly a cause and partly an effect of quite unreasonable expectations on the part of the users.

The biggest arguments continued to relate to depth of water, as they had from the 1860s onwards. Depth, it must be stressed again, governed not only the maximum size of ship which could use a particular dock or entrance but also how much freedom to come and go at a variety of states of the tide an entrance could provide for smaller vessels. Depth of water was the most expensive attribute of any dock, and it was a usual engineering yardstick that excavation and construction costs varied roughly as the cube of the depth. It will be clear from earlier chapters that depth was still very much a live issue in the period covered by this book, as three of the four major construction schemes were concerned largely with providing enhanced depth. This issue, and the position of the shipowners, will be considered in due course, but first this chapter will address some of the smaller issues which, taken together, nevertheless had a considerable bearing on the overall effectiveness of the port in shifting goods – and this is perhaps the place to remind the reader of the old truism that docks are not built for ships, they are built for goods.

During the Great War, Liverpool's docks became a national problem when crucial war supplies were delayed by acute congestion which could leave inward-bound ships waiting many days anchored in the river until a berth became free. Port authorities have long "benefited" from the advice of complete amateurs who can solve their problems in some amazingly simple way. It was obvious what was wrong at Liverpool: there was a shortage of dock labour, and what there was had been depleted in quality as well as quantity by the loss of those who had volunteered for, or later been conscripted into, the armed forces. This under-sized pool contained a significant proportion of older men and, of course, everything would be much improved if a way could be found to make them pull their weight.[1]

But while the war clearly exacerbated congestion, it did not initiate it. The shipping industry is acutely cyclical. Trade had been improving rapidly since 1911, and 1913 had been a particularly good year which had resulted in acute congestion on the quays; 1915, when important people like Lloyd George were getting involved, established a new record for a year's tonnage of shipping using the port. Liverpool exhibited the slightly strange phenomenon that, other things being equal, a graph of the cargo efficiency of any particular dock expressed as tons of goods per yard of quay per year normally closely parallels a graph of the total throughput for the dock and broadly echoes that for the port as a whole. This suggests that congestion was not necessarily as bad as we have been told, that it was a serious annoyance rather than a disaster. Otherwise, efficiency would have begun to level off as the aggregate throughput continued to rise, followed by a saturation point beyond which the throughput for a particular dock could not rise.

But the *Annual Accounts* show a slight but definite fall in the receipts of town and dock dues, both of which were charged on the nature and tonnage of the goods rather than on the ship. Resort to the more detailed Dock Revenue Statements reveals the reason: the high-earning docks almost without exception show an increase in dues on inward goods but a decrease on outward ones. This means that there were, as alleged, more ships leaving in ballast since the outbreak of war, possibly as a result of a shortage of labour to load them when the more pressing need was to clear the berths for further import cargoes. If that was so, then the export goods must have been stopped short of the Dock Estate, for the notorious narrowness of the quays, sheds and avenues of the older docks could not have accommodated them, and such was indeed the case, resulting in congestion in railway yards and a shortage of wagons, sheets, and the like. Even small horse-drawn carts suffered acute delay through avenues being blocked by goods which had been piled outside sheds. It was claimed that inward-bound ships might spend days at anchor in the river waiting for a berth.

There were, of course, those ready to attack the Board for causing this state of affairs by refusing to invest in improved facilities, chiefly in two specific areas. The first was the raising of the old cry about the need for increased crane facilities, which had been popularly blamed for quayside congestion during previous "highs" in the trade cycle. In 1915, as in 1912 and 1872, the problem was not getting goods into or out of ships; it was completing the paperwork (weighing, tallying, etc.) and getting the goods to and from the quayside. More

---

[1]This aspect of the problem is considered in a more judicious manner in K.R. Grieves, "The Liverpool Dock Battalion: Military Intervention in the Mersey Docks, 1915-1918," *Transactions of the Historic Society of Lancashire and Cheshire*, CXXXI (1981), 140-158.

or faster cranes would have served only to exacerbate the chaos in the sheds and on the quaysides unless they were part of a larger modernisation involving additional storys on sheds.

The second criticism levelled was legitimate – to a degree. The quays and avenues of Liverpool's docks had always been penny-pinching in their use of land because they needed to be built on that long narrow strip of "land and strand" sandwiched between deep water and expensive land. Extension beyond the minimum in either direction resulted in rocketing costs per acre.[2] It is also perhaps worth mentioning that building docks with a vestibule and branches (like Alexandra) appears to offer a better quay length-to-water area ratio than an old, more-or-less square, sailing ship dock (like Albert) and hence a more effective use of the total area. So it would if we were comparing like with like, but we are not, for the E-shaped docks were built for much bigger ships and therefore needed more "swinging space" for manoeuvring them in and out of the branches. It may be counter-intuitive, but Gladstone's quay-to-water area ratio is actually about half that of Albert, an archetypal square dock built for sailing ships.

As the cargo density of ships increased, so quaysides which had always been meanly proportioned became increasingly inadequate for the burdens placed on them. The main solution was to build upwards: the first fully-developed[3] double-story transit sheds were those at Harrington Dock (1883) and the first three-story ones were completed at Huskisson in 1894. Those at Harrington cost considerably more per square foot than single-story sheds, but they really were capable of discharging a vessel at twice the speed. It is true that the Board had invested in expensive high-technology sheds, and it is also true that it would have been better if it could have invested in more of them, but the reader will recall

---

[2]It is admittedly an extreme case, but the Somerset Dry Dock (Malta) shows something of the costs of going into deep water: it cost, in round figures, £150,000, of which no less than £20,000 was spent on coal for pumping the site. The last twenty-three percent of its depth accounted for ninety-four percent of the pumping costs. C. Andrews, "The Somerset Dock at Malta" *Minutes of the Proceedings of the Institution of Civil Engineers*, XXXIII (1871-1872), 352-375.

[3]There are problems of definition here: a shed built at Dukes Dock in 1857 (not then part of the Mersey Docks and Harbour Board [MD&HB] Estate) had an upper floor which was at about two-thirds the height of the roof valley, and the "wrecked goods depot" at Sandon (1858) also had an upper floor. Both buildings were out of the usual run of things, the former because it was built for barge traffic and the latter because it was not strictly a transit shed. After the manner of the *Engineer's Reports*, which tend to have their own private meanings and expressions, we find no mention of two-story sheds until many years after what ordinary people would call a two-story shed had been in service. The distinction is that those at Harrington were designed and equipped to discharge goods direct from the ship's hold to both floors simultaneously.

from other places in this book that there were many calls on the Board's funds; it had to prioritise, and at this time it had made the completion of Gladstone its top priority. Had war not broken out, one would have praised it for its resolution in following the right long-term policy without being knocked off course by the howls of protest brought about by an upturn in trade.

Some wartime activities, however necessary, were themselves disruptive to traffic. When *Aquitania* returned to Liverpool in 1919 for refitting for peacetime use, "a large portion" of its fittings were in the shed to the north of the Gladstone Graving Dock, and the fact emerged that most of that new shed, at 900 by 100 feet among the biggest in the port, with good railway connections to both front and rear, standing alongside the 1000-foot deepwater berth, had spent most of the war filled with furniture and fittings from it and other liners. Repeated dry-dockings of major liners in government service and warships largely prevented the use of the dock for cargoes anyway, and its earnings in dock and town dues rank it down among the less successful coastwise docks.

Good Housekeeping

This brings us to the subject of ineffective use of space and general clutter, a factor in port efficiency which has not received the attention it deserves: historians have tended to look at a plan of whatever port is in question and assume that "what you see is what you get." In the period which concerns us, it very definitely was not.

The Engineer's Department was bound to accumulate rubbish and equally, in an organisation which always viewed itself as "penny wise," any rubbish that could be re-cycled would be stacked up until there was enough to make it worthwhile recycling it in-house or selling it. Thus, for example, the published accounts of the old body of Dock Trustees for 1827 have entries for £616 received for "old iron and ropes" and another £929 for "old building materials." Lacking as we do an historical price index for unspecified old building materials, we must rest content with the assumption that £929 represented a very large amount. At the time, good quality new sandstone cost about sixpence per cubic foot - about 7s 6d per ton - so we are clearly looking at a heap of thousands rather than hundreds of tons of materials. As the Engineer's Department and its work became ever larger and more complex, it was logical that it would generate more rubbish, but one might also expect that having a much-enhanced office staff (including a substantial accounts section), it might be better able to balance income from scrap materials against the cost in space occupied. At least, one might expect that if one had no experience of the history of the Dock Estate.

Sales by tender of old materials from the Dock Yard were normally held once a year, and the 250 tons of mixed iron and steel scrap, fifty tons of scrap chain, ten tons of old rope, twenty tons of broken glass and much more besides sold in 1893 was apparently just one year's output of rubbish. The space occupied by all this was obviously considerable, and in a number of years we find machine-shop ferrous swarf - turnings, borings, etc. - reaching fifty tons, which would occupy very roughly forty cubic yards. This situation changed little over time, except that in the later stages of the war a shortage of brass for shell-cases resulted in a government drive for scrap brass. In 1916, among the materials sold from the Dock Yard we find fifty tons of turnings and eighty tons of old firebars. There was an overspill yard for scrap on the west side of Brunswick from which the 1915 disposals included 170 tons of wrought iron scrap, sixteen tons of condemned chains, fifty tons of old dredger buckets, 110 tons of railway metal and eight tons of carbide drums. This overspill in its turn spilled onto the adjacent quay, with a small marine boiler, thirty tons of corrugated iron sheeting and more.

The reason for all the tedious lists of rubbish above should become clear. It was not just a question of the Engineer not minding if the Dock Yard was untidy or cluttered. Brunswick had been extensively rebuilt as part of the works under the 1898 Dock Act, completed in 1908. It provided some of the best and most modern berths in the port - when they were not covered in heaps of scrap corrugated iron or the forty tons of old timber which lay at the southwest corner of the dock.

This intrusion on the trading area of the port was not confined to the parts in close proximity to the Dock Yard, but spread all over the Estate, literally from one end to the other. In 1915 we find quantities of carbide drums at Gladstone and the east side of Hornby, eighty tons of railway metal at Canada No. 3 Branch, and ten tons of oak staves at the northwest corner of Canada. The following year there was "an old fractured cast iron cylinder weighing about 12½ tons" at Herculaneum, and in 1917 there were eighteen tons of broken glass at Alexandra No. 3, six tons of old rope at Canada No. 3 and, to help the war effort, "1400 brass tops of electric lamps." In 1918, the Engineer reported that he had "sold 157 tons of Driftwood lying at various points on the Dock Estate at 15 shillings per ton."[4]

The Engineer also had three waste depots, whose main mission was to receive and sort quay sweepings. The horse manure was separated out and heaped up until there was a canal boat-full of it, which was then sent to the agricultural area north of Liverpool, while that for which nobody could think of any use whatever was loaded into a hopper barge and dumped at sea. Each of the three

---

[4]Driftwood in this context means wood which had recently fallen in the water and which the original owner could not be bothered to recover. The Board had to remove it anyway as a hazard to navigation, but it could be quite valuable stuff.

depots, therefore, had a 120-foot berth reserved for the hopper barges. The scale of operation was considerable, involving three Yorkshire steam lorries.[5]

Other departments also contributed smaller amounts of clutter. The Warehouse Department, for example, used to store ash from the burning of condemned tobacco until there were several tons and then sell it to a fertiliser manufacturer. Also piled up to await sale were floor sweepings from warehoused corn, tea and tobacco, and considerable quantities of scrap wood. It was normal to sell old bale bands in lots of not less than half a ton, those too being heaped up until one of the "rummage sales" occurred. Because of the delays on the quaysides, there was extra pressure on warehouse space and some of that, too, was wasted. The Marine Surveyor's workshops at Herculaneum Buoy Store seem to have worked in a different manner, for their scrap, which must have been plentiful because the workshops' staple business was repairing buoys, hardly ever features in the invitations to tender.

The Board was perfectly well aware that it was wasting valuable space. In 1913 a Special Committee recommended that one of the General Manager's staff should go, preferably accompanied by one or two Board members, to inspect the various yards to ensure that there was no hoarding of rubbish. Before that, it had given the General Manager powers to hold rummage sales when he saw fit, without reference back to the Committee, and in 1916 it was letting contracts to scrap merchants which involved their coming to collect when summoned. One could quite think that, like many things the Board did, it was a bit slow but it worked out right in the end. Alas, it was not so. In 1928, the Harbour Master was asked to report on a request for some extra bollards at the White Star Line's berth at West Gladstone. He agreed they would be useful, but "as the quay is at present occupied by gear, gravel etc belonging to the Engineer's Department, the consideration of this matter might be deferred until the Engineer's work will permit of the quay being cleared of this material." The berth in question was, of course, one of the best in the port.[6]

None of the above should be taken to imply that all the guilt for clutter lay with the Board, since it certainly did not. Shipowners who undertook their own master porterage would discharge goods anywhere and in any manner if nobody prevented them. In March 1912, for example, J. Bibby and Sons, Ross

---

[5]The history of quay sweeping in Liverpool is complex and may be followed in National Museums Liverpool (NML), Maritime Archives and Library (MAL), MD&HB, Worked-up Papers (WUP) 207/1.

[6]There are a series of WUPs and Unbound Worked-up Papers (UWUP) that deal with the issue of the accumulation of rubbish in the docks. See, for example, NML, MAL, MD&HB, UWUP O14, vol. 3; UWUP C169; WUP 58/1, WUP 58/2; WUP 141/1; and WUP 162/1. The quote is from Works Committee, minutes, 27 January 1928.

T. Smyth and Co. and William Harper and Sons all had the problem of their vessels arriving at a berth only to find that its quay and shed space was still full of goods from the previous vessel discharged. This would come as no surprise since in January Mr. Moss had informed members that "today, there is not a clear berth in Liverpool, while there are three foul berths at the North End and six at the South End."[7] The issue that the Board was discussing was that of "Special Rents." Normal port charges included the right to leave goods on the quay or in the sheds for up to forty-eight hours free of further charge. After that, the Board was empowered not only to charge very high penalty rents but also to waive them in favour of a purely nominal "special rent" (often of one farthing per square yard per day), and the situation had developed whereby anyone with an even half-decent excuse was likely to be granted the special rate. The discussion became quite heated at times, particularly in clashes between Mr. Rome and Mr. Hodgson, which are more readily understood when it is known that the former was a shipowner and the latter a provision merchant.

There was no real answer to the question: consignees granted the special rate would not make much effort to remove their goods, while those charged the penalty rent might well find that by the time they got around to collecting the goods, the rent exceeded the value. In that case they would try to get the Board to relent, but in the last resort would simply abandon the goods where they lay. If consignees could be difficult, so could shipowners: Worked-up Paper 120/1 is entitled "Wrecks in the Docks," and much of its contents relates to vessels which sank in dock and were abandoned because they owed more in the cost of raising than they were worth. Most of the offending vessels were small – flats or barges – but in one sense these were more trouble than larger vessels because they sank out of sight, making it necessary to buoy, light and watch them until they could be raised. In 1920 (to take a random example) at least twenty sank in dock.

Space was also long occupied by the 120 by sixty foot "Wrecked Goods" shed at Sandon, mentioned above. Confusingly, this did not normally mean goods salvaged from vessels which had sunk in dock. When it was built it was to receive the cargoes or part-cargoes of "vessels returned in distress," i.e., those which had left port in cargo and then suffered damage sufficient that they had to return rapidly to the nearest graving dock – which they could not enter until they had discharged their cargo. By the period with which we are concerned this was a comparatively unusual occurrence, but it serves as a reminder of yet another perhaps unexpected source of congestion on the quaysides.

---

[7] *Ibid.*, Discussions at Board, 25 January 1912.

It was also common for ships to leave quantities of "old gear" obstructing the quaysides.[8] The tenants of appropriated berths were allowed to have "ash bins," which were in fact for any form of dry waste, and these bins were limited to a size of twelve by six by four feet deep, or five percent of the superficial area rented, whichever was the greater. They may have helped keep the working surface of the quays clear, but they were an encumbrance in or behind the shed. Perhaps most surprisingly the tenants could leave assorted ships' gear fouling their own working space. In 1913 the Board found that some 16,137 square yards of shed space was thus occupied in the docks from Hornby to Huskisson inclusive, and that the White Star berths at Canada (among the most desirable in the port) provided no less than 5520 square yards of that total.[9] As one got to rather less up-market firms in the docks further south, the amount of waste space from this particular cause diminished to perfectly acceptable proportions. Then there were huts. Huts were the Traffic Department's nightmare: they were fairly small (typically about eight by six feet), mounted on small iron wheels and used as temporary offices or equipment stores. Despite the Board's long-standing and resolute attempts to control them, the proliferation of huts was still a problem in the twilight years of break-bulk.

If a combination of the Board's bad working practices and the anti-social behaviour of the customers meant that there was less quayside available at any one time than there should have been, there was also less water space. Quite apart from the abandoned vessels, those under arrest and the occasional one sitting on the bottom, we must take account of the mooring requirements of the Board's own vessels, which constituted a "fleet" larger in aggregate length and tonnage than those of many of its customers.

In 1920, for example, the Engineer's Department had the largest number of vessels, with five sand pump dredgers, including the largest in the world,[10] four centre ladder dredgers, two double-side ladder dredgers, seven grab dredgers, fifteen steam hopper barges, five floating cranes with capacities up to 200 tons, a tug, a survey tender, and eleven "flats." But that was by no means the end of the story: the Marine Department had four survey launches, two steam tenders, two steam yachts, and a River Police launch. Because the Board, rather

---

[8]On 10 December 1913, for example, the Docks and Quays Committee discussed the Special Committee Report mentioned above, and it was reported that "the Chief Traffic manager had already taken steps to curtail, as far as possible, the space occupied by ships' gear, propellers, tail shafts, etc. inside and outside the sheds at appropriated berths." NML, MAL, Special Committee, 1911-1913, minutes, 19 October 1913.

[9]*Ibid.*

[10]Namely *Leviathan*, of 10,000 tons capacity; see chapter six.

than Trinity House, was responsible for lighting the approaches to the Mersey, there were five light vessels, of which four were on station, but the fifth, held in reserve to be deployed in case of accident, required virtually permanent moorings. The Marine Department also owned substantial salvage equipment, consisting of two "watch vessels," four motor launches, a salvage steamer, and seven assorted lumps and camels.[11] Then came the Pilotage Department, with four steam pilot "boats" (which were actually small ships) and a launch, the Warehouse Department with eight grain barges and finally the Harbourmaster's Department with two tugs.[12]

This fleet had been built up over a long period of time to respond to differing (and growing) needs. The mooring space required by the vessels was not as great as might at first appear, since the really big ones (the sand pump dredgers) spent most of their time out in the river and approach channels, and many of the small ones could be moored several deep in some obsolete dock like Stanley. But the law of the conservation of mass indicates that they all had to be somewhere: since the modernisation of Brunswick, the old Brunswick Half-Tide had been effectively closed for commercial use and became a home for many of them, and that alone represented about one percent of the entire water area of the dock system.

The question of dredging is discussed elsewhere, but for present purposes it is necessary to note that the grab dredgers were used only in the docks and around the entrances, that is to say that at any time they were in use they were obstructing berths or access to them. Conversely, congestion made dredging difficult: when the Harbour Master (who had to field the shipowners' frequent complaints about inadequate depths at the berths) sent a slightly irritable memo to the Engineer-in Chief stating that "difficulty and inconvenience is frequently experienced owing to the reduced depth of water available," he was only doing his job. But when a couple of years later *Cedric* grounded in Huskisson Branch, Newell pointed out quite bluntly that he could not dredge the dock because there were too many ships in it.[13] It is just another example of the conflict between different needs which is such a widespread characteristic of the history of ports.

---

[11]Lumps and camels were vessels constructed or adapted to be partially sunk, secured to a wreck and then pumped out to raise it. In the river they could sometimes be floated over a casualty, secured to it at low water, and the large tidal range would do the rest, allowing the wreck to be safely beached. See NML, MAL, MD&HB, ENG/3, Engineer's Commonplace Book No. 3 (1889-1899), 63.

[12]*Ibid.*, MD&HB, *Yearbook*, 1920.

[13]*Ibid.*, Works Committee, minutes, 27 October 1916 and 13 October 1918, respectively.

What does seem strange, however, is that earlier that year Newell's men *had* managed to get to another appropriated berth of the Oceanic Steam Navigation Co. in Canada No. 1 in order to do some dredging. On 17 April the Harbour Master sent a memo informing the Docks and Quays Committee that the grab dredger *Miles K. Burton* had been much delayed and slightly damaged by having to handle debris which included "iron, ropes, hand trucks, wire etc.," and this had apparently happened on several previous occasions. Here we have a major shipping line operating large, deep vessels allowing their own in-house master-porterage team to foul a berth that was rented for its exclusive use – and whose manager, Colonel Henry Concanon, would next year be elected a member of the Board. Lest this be thought a quite peculiar degree of misbehaviour perpetrated at a time of acute national need for quick turnarounds, it should be pointed out that Elder Dempster was no better, leaving its berths at Toxteth and Harrington severely fouled: in one case "the whole berth was found to be a mass of old rope, wire, dunnage mats and cargo sweepings," while in another there was "a large amount of rope, wire, dunnage mats etc."[14] The Board may have had its failings, but not all causes of delay in ports were under its control. If the Board's employees had a culture of poor housekeeping which reduced the efficiency of the port, so too did their customers.

The annual amount of dredgings from the docks (Liverpool and Birkenhead together) during the war years was typically a little short of 750,000 tons, roughly 1500 hopper barges full.[15] The 3000 "ship movements" at the dock entrances this represented was equivalent to over five percent of the number of cargo ships arriving and departing the port in the congested years of 1912-1915.

This cannot be a precise econometric exercise. The comparison above, for example, is potentially misleading, since the true total of "vessel movements," including tugs, forwarding barges, ferries, passenger tenders, etc. is incalculable but is several times the apparent total. One little window of numbers exists: in 1925 the Harbour Master enumerated the vessels other than ships which used the river entrances in the four days ended at midnight on 7 June. There were two floating elevators in and two out, three floating cranes in and three out, sixty-three hoppers and dredgers in and fifty-nine out, 311 tugs in and 324 out, and 483 flats and barges in and 506 out. This is a total of 1756 vessel movements, 439 per

---

[14]*Ibid.*, UWUP, M63, memos from the Engineer-in-Chief to Docks and Quays Committee, minuted 3 May 1922 and 9 May 1923, respectively.

[15]*Reports of the Acting Conservator*, various years, give figures for annual amounts dredged.

twenty-four hours.[16] To those must be added the movements which passed unrecorded through not involving the use of a river entrance, which in the case of the barges would be a considerable addition. It seems safe to make an educated guess that something of the order of five percent of the port's berth space was unavailable to shipping at any one time owing to the prodigal use of scarce space by both the Board and its customers, and the extensive necessary operations of the Board's own vessels.

Figure 5
Building the New 3-story Ferro-Concrete Transit Sheds at Brocklebank in 1907

Source: National Museums Liverpool, Photography Department, Ref. No. N86-271.

[16]NML, MAL, MD&HB, Legal B74, draft proof of evidence of Richard Holt before the Select Committee on the Mersey Tunnel Bill, 1925. Further information on barge movements is to be found in chapter seven.

In addition, maintenance and minor quasi-capital construction need to be taken into account. In 1913, for example, among other such works, extensive improvements to sheds at Carriers, Toxteth and Alexandra docks were in hand. These were not all completed by 1915, when shed works at Nelson, Huskisson and Brocklebank were begun.[17] Each of these projects would put at least one berth out of action for a minimum of several months. Some approximate measurements made by colouring in affected sites on maps of the Estate suggest that at any one time roughly half to perhaps one percent of the port's berthing were in the hands of the Engineer for comparatively minor works of this type.[18]

Major capital works could be a berthing disaster: between 1873 and 1908 something approaching ten percent of the port's berths were unavailable from this cause at any one time.[19] Fortunately, this was not a problem during the period of acute congestion which concerns us here because virtually all available resources were applied to the completion of Gladstone, causing minimal disruption to other docks. But such was the scale of the operations at Gladstone that what might seem minor disruptions were still fairly large, as related in chapter two, when Newell asked for some shed space to be handed over to him temporarily to store cement - in the shape of half the West Hornby shed, 600-feet long and 100-feet wide.[20] Six months later a Works Committee inspection found the shed "practically empty," but when the cement finally did start to arrive in October the first delivery amounted to 1200 tons.

By a similarly happy coincidence, the extensive Central Docks modernisation works described in chapter three, which at other times would have caused absolute chaos in the coasting trade, were carried out when trade was slack and alternative berths relatively easy to find.[21] In short, the congestion problem which occurred between, say, 1912 and 1920, could have been very much worse. Had the Board decided back in 1904 not to build Gladstone but to undertake a major modernisation of Hornby and Alexandra, after the manner of

---

[17]Works such as these are conveniently listed in the "Unappropriated Receipts Account" published in *ibid.*, *Annual Accounts*, various years.

[18]This an admittedly imprecise method, but because it takes no account of the Engineer's equipment and materials encroaching on adjoining berths, which we know they did, if it errs it does so on the low side.

[19]Estimated in the same manner as described above.

[20]NML, MAL, MD&HB, Docks and Quays Committee, minutes, 8 January 1919.

[21]This was, of course, another benefit of the counter-cyclical investment which had been something of a Liverpool trademark since the days of Jesse Hartley.

the Canada-Huskisson modernisation of 1891-1901, the congestion, especially between 1912 and 1920, would have been catastrophic.

It would have been perfectly possible to make an improvement of a small percentage in the overall efficiency of the port at virtually no expense by the introduction of a ruthless programme of getting rid of junk on a weekly basis, and the Board's statutory powers were quite sufficient to enact a new by-law allowing it to seize any ships' gear left on the quayside for more than forty-eight hours and charge heavily for the dredging of appropriated berths if the grabs brought up ships' refuse. The fact is that neither the Board nor the customers really took the problem of rubbish seriously, resulting in diminished efficiency and a great deal of inconvenience.[22] The quay and shed space in the port was phenomenally valuable and could not be readily increased. That the Board should have allowed customers to occupy up to five percent of the appropriated berths with "ash bins" is surprising; that shipowners should have wished to do so beggars belief.

The Problems of Debt and Depreciation

The finances of the Board at the time of the decision to build Gladstone Dock are mentioned briefly at the end of chapter one. This section expands on that and considers the way in which the Board's customers viewed the improvement of port services and the means of paying for them, which might be summarised as, respectively, "yes, please" and "no, thank you."

The fundamental points crop up in various parts of this book. Because of its physical limitations Liverpool was among the most expensively-engineered ports in the world, and the docks had been built entirely on borrowed money for whose repayment and interest costs there was virtually no income except from the users of the port through charges on ships and cargoes and the rents from land or buildings.[23] The result, entirely predictably, was that the Board's debt rose inexorably which, other things being equal, would mean that each new capital project would be more difficult to fund than the last.

For the whole of the nineteenth century, other things were not equal: the port existed in a continuum of growing trade which enabled its revenues to keep pace with, and sometimes overtake for a while, the demands of servicing its

---

[22]Oral evidence suggests that other businesses having no direct connection with the docks would fly-tip rubbish on the Estate because no-one would notice a bit more.

[23]The main exceptions are in the rentals of land which had been purchased for possible future expansion and for which the Board had no immediate use, including much of what later became the site of the Tranmere Oil Terminal. For further details see A. Jarvis, "Land Policies in the Port of Liverpool, 1857-1930," *International Journal of Maritime History*, XIV, No. 1 (June 2002), 115-133.

debts. Annual growth slackened after about 1860, but as shown in chapter one, simple percentage growth remained impressive. This growth could have extremely beneficial effects; it is clear that at times Jesse Hartley and/or the Sub-Committee of Works, to which he reported, underestimated the rate of growth in the width of paddle steamers; Wellington and Sandon Docks, opened 1850 and 1851, respectively, had the widest entrances in the port but were soon outgrown. Happily, such was the growth in trade that more new docks were urgently needed anyway and Canada (opened 1858) was built with a 100-foot entrance, not exceeded until Gladstone was opened. By 1888, when the works under the 1873 Act were completed, the port did not need any further major extensions; it needed improvements.

Here lay the rub: obviously a modernised old dock would handle more tonnage and earn more revenue than an unmodernised one, but it is clear that quite often the effect was that a few favoured customers moved from their existing berths to better ones, thus allowing everybody lower in the pecking order to step up one rung, resulting in some docks held in particular disfavour being left with little or no trade. The severity of Albert's decline after the completion of the works under the 1898 Act is a conspicuous example.[24] In short, the demand for major construction works, which before about 1850 was largely quantitative, had become at least partly qualitative. The growth in tonnage of goods handled continued, but it did not increase in a manner commensurate with construction and modernisation. Liverpool thus found itself on the treadmill familiar in many ports of having to spend ever-increasing sums to keep its existing water area up to date in order even to maintain, never mind increase, its trade. When the 1898 programme was eventually completed in 1908, it had increased the water area of the docks, especially at the south end, but not to any great extent; the "new" Kings Dock, for example, took in the extensive site of the old tobacco warehouse, but owing to the need for space to swing longer ships around, was only a little over an acre larger than the old one. What had cost the money was providing deeper water, wider two-story sheds, and better railway access to the quays. This was a question of keeping up with functional depreciation.

Kings was a very old dock, opened in 1788, and it might be expected that after a century it would need major redesign. It had, however, been somewhat enlarged and improved by John Foster in about 1810, by Jesse Hartley in about 1850, and in 1893 it had benefited from the installation of impounding pumps at Coburg; it might have been old-fashioned but it had not lain untouched. The Sandon programme, which lasted from 1891 until 1901, increased the water area only by about eleven acres – so that if one views it strictly as an extension

---

[24]For the long decline of Albert, see M.K. Stammers, "A Dock Where Nothing Will Ever Be Done: Albert Dock in the Twentieth Century," in A. Jarvis and K. Smith (eds.), *Albert Dock: Trade and Technology* (Liverpool, 1999).

it cost some £20,000 per acre – this at a time when the "ball-park estimate" for a typical new dock would be about £20,000-£25,000 per acre.[25] But the improvements at Sandon were designed to "unlock" a much larger area of existing docks by deepening works and constructing a new river entrance and, depending on exactly which other areas one includes, the figure falls considerably, but the costs of accommodating the North Atlantic greyhounds remained high.

It gets worse than that: The New North Works component of the 1873 programme, built to take the exceptional-size vessels, had cost in round figures £2,500,000 and within less than ten years of their completion needed replacement in that role by the new works at Sandon. They had, in effect, already depreciated by £1,600,000 minus the value of the extra eleven acres gained – of which £300,000 would be a generous estimate. Even that eleven acres was not outright gain, for it increased the water space in relation to the quay space, necessitating the construction of a three-story transit shed in 1908 to compensate for the loss.

Fortunately, that probably marks the worst of it, and the potentially appalling effects of such problems were mitigated by low interest rates, so that the percentage of the Board's dues income spent on interest remained remarkably stable until the turn of the century (the situation then is described in chapter one).

This has been a lengthy introduction to a curious fact, which is that when shipowners discussed the levels of port dues in Liverpool or elsewhere they always expected port authorities to budget to break even without making allowances for depreciation of hugely expensive fixed assets and without acknowledging the ongoing cost of servicing debt. The Board took a longer-term and more responsible view, in part because of the recognition, referred to in chapter one, of the importance of appearing creditworthy. Unfortunately, there were two popular means of "measuring" the soundness of a port's finances which were not just inadequate but downright misleading. The first was the relationship between the statutory maximum rates of dues and the rates actually charged: the greater the difference the better the situation of the port authority. This would be fine in a situation where ports were able to suppress competition, as they were when widely separated or under the control of a government or even a cartel, but these conditions did not apply in Britain, and ports, like shipowners, could only charge what the market would bear.[26] What they were allowed to charge is

---

[25]This should not be taken to imply that any Liverpool docks were actually built for that price – this was just the very crude method the engineers employed when giving a preliminary idea of order of cost.

[26]There were, however, possibly valid accusations that some of the railway companies achieved local monopoly conditions by their ownership of several ports, such as the Great Western Railway with its South Wales ports.

completely irrelevant. Just as misleading is the second "measure," which is the difference between the borrowing powers granted by Parliament and the amount of money actually borrowed. The larger the difference, the sounder the port's finances. This makes the assumptions that Parliament in its wisdom would never grant borrowing powers in excess of what the port authority could afford and that rates of interest would not change significantly between the granting and the last exercise of those powers, which might be ten or fifteen years - or in the case of Gladstone, twenty-one years. Neither assumption, of course, is remotely tenable.

The Board's debt rose inexorably throughout the period, but the proportion of dues income spent on interest varied approximately inversely with the tonnage of shipping using the port, showing falls in the 1911-1913 and 1919-1921 booms and during the recovery from the Great Depression. This could be coincidence, but it could also be an indication that the Board's judgement of how much it could afford to borrow and when was much more astute than we might expect. If so, it got little credit for it from the shipowners.

The last years before the Great War were marked by a continuation of the shipowners' policy of crying for the moon while not wishing to pay. This was particularly marked in the case of Cunard and the Gladstone Graving Dock, detailed in chapter three, but there was a slight but perceptible weakening in the shipowners' position as compared with that which obtained earlier. Some of this may be seen as a loss of respect or "moral authority" stemming from several widely different causes. The "sell-out" of the iconic White Star Line to the American International Merchant Marine (IMM) had an impact which is difficult to appreciate in our unpatriotic times. This was a complex matter, but it could be, and was, easily simplified for the general public.[27] Even to a more sophisticated commercial readership, the IMM was still normally described as "J.P. Morgan's IMM," emphasising the part of the financiers rather than of the shipping men. The IMM paid very inflated prices for much of its tonnage, and the stock was heavily "watered," with the result that while the commission-takers did well, the ordinary shareholders fared badly. When Bruce Ismay became President this might have been seen as a gain to British shipping; instead, it was seen at least by some as a further betrayal.[28]

---

[27]For example, "when it became known that the White star Line was to be sold to the American group, a public sensation followed what was regarded as a blow directed at the foundations of the British mercantile marine." C.E. Lee, *The Blue Riband. The Romance of the Atlantic Ferry* (London, n.d. [1930]), 182. For a detailed account of the subject see W.H. Flayhart, *The American Line, 1871-1902* (New York, 2000), chapters 16 and 17.

[28]B.W. Ginsburg, for example, got close to the boundaries of contemporary courtesy when giving a paper entitled "Shipping Combinations" at the British Association of the Advancement of Science meeting in Belfast - Lord Pirrie's home town - in 1902:

The old image of the patrician shipowner took another beating with the Report of the Royal Commission on Shipping Rings, published in 1909. The evidence was complex, and it is by no means necessary on reading it to come to the conclusion that shipowners were monopolists or profiteers, but to some it could and did appear so. It is arguable that a half-way house - suppressed competition - between monopoly and untrammelled competition was exactly the way to provide shipping services at a price which was both affordable and sufficient to allow for new investment, including, through the payment of realistic port dues, investment in ports. Manufacturers in particular, however, would not look beyond the doctrine of competition producing the benefit of cheap freight rates.[29] When such a view was published, the rapidly-increasing trappings of success visible among shipowners became a positive disadvantage in terms of image outside the shipping industry.

The loss of *Titanic* was a serious blow to the image of shipowners, but it was rendered worse by Bruce Ismay taking a place in a lifeboat. This was the age of admiration for British heroes in the mould of the men who went down with *Birkenhead* - and Ismay's action made him appear to some a coward as well as a traitor. Had he, like the Captain, gone down with the ship, the subsequent enquiries in Britain and America might have passed more easily. There was little direct personal condemnation of Ismay, and the specific accusation that he had encouraged or even instructed the Captain to proceed at a reckless speed was not proven. The passenger liner industry as a whole was, however, dealt with fairly harshly in both enquiries.

Then came war. It is interesting to note the shift in editorial comment on the shipping industry in *The Economist*, which started out, as one might expect, praising the efforts of shipping companies in keeping vital supplies moving under conditions of acute difficulty. But in 1915, Bonar Law invested £8110 in shipping shares and within two years had "earned" a profit of £7471: by the time *The Economist* came to comment on this its tone was very decidedly changed.[30] In 1916 the Cairn Line paid a dividend of ten percent - and a bonus of thirty percent. *Fairplay's Annual Summary of Shipping and Finance* stated that even

---

"Mr. Morgan himself is to receive some 2½ million sterling [*sic*] and doubtless other persons who had contributed towards this huge amalgamation would likewise obtain their reward." Reported in *Engineering*, 21 November 1902, 662-663. According to Gordon H. Boyce, *Information, Mediation and Institutional Development: The Rise of Large-Scale Enterprise in British Shipping, 1870-1919* (Manchester, 1995), 104, "Pirrie and Ismay were viewed as pariahs..."

[29]The comments on the Report in *Engineering*, 4 June 1909, 761-762, for example, treat this as axiomatic.

[30]*The Economist*, 7 July 1917.

after payment of Excess Profits Tax, the five-year average dividend on ordinary stocks in liner firms was 11.26 percent, compared with 6.83 percent immediately before the war, when trade was very strong. These were the firms which, when peace came, complained bitterly that port dues were driving them to bankruptcy. Even firms which enjoyed a very high reputation for ethical behaviour, like Blue Funnel, made huge profits almost by accident.[31]

For some, including Lord Vestey, the war brought not only large profits but also the opportunity to invest some of it in a donation to the Liberal Party, which sometimes seemed connected with the receipt of a peerage or a baronetcy.[32] The subsequent Parliamentary Enquiry into the sale of honours was a total whitewash - and widely recognised as such. Then came the scandals surrounding Lords Inchcape (of P&O) and Kylsant (of Royal Mail), the latter actually being imprisoned for issuing a false prospectus. These were not good times for the prestige of shipping magnates, and from newspaper accounts one senses a feeling abroad that things were not what they used to be, which was possibly true. The Kylsant collapse was among the greatest ever in Britain, and one of the companies dragged down in the process was Elder Dempster. One suspects that Alfred Jones was too honest, too clever, or both, to preside over the dubious accounting methods used by the Royal Mail Lines - or at least to get caught doing it.[33] It seems, too, that with the passing of his generation, control of major shipping lines passed from the hands of the innovators and entrepreneurs to the professional managers. This was not, of course, an entirely correct perception, and even if it were correct it was not rational, but the stereotype of the "great man" was strong at the time and it was one into which Alfred Holt, Thomas Ismay and Alfred

---

[31]*Fairplay's Annual Summary of Shipping and Finance* (London, 1928), which contains a retrospective account of various figures.

[32]For reasons why the Vesteys were unpopular, see P. Knightley, *The Rise and Fall of the House of Vestey. The True Story of How Britain's Richest Family Beat the Taxman - and Came to Grief* (London, 1993).

[33]This is not necessarily a fair comparison; see P.N. Davies, "Group Enterprise: Strength and Hazards. Business History and the Teaching of Business Management," in S. Marriner (ed.), *Business and Businessmen: Studies in Business, Economic and Accounting History* (Liverpool, 1978), 141-168. Jones' success came in the continuum of growth mentioned above, while Kylsant's failure came at a time of particularly difficult trading conditions, and Davies argues that, even so, had he managed to hang on a little longer the group might well have recovered. But that, of course, was not how outsiders perceived it at the time.

Jones, to name but three, fitted.[34] Men of the newer generation, even if of great ability, in the main did not, though there were obvious exceptions like Sir Richard Holt or Sir Alfred Read.[35] The increasing influence of trade associations, touched upon above, was probably a contributory factor here as well, albeit only a minor one.

None of this should be taken to mean that requests for reductions in dues were being made by an industry which was thoroughly profitable and well able to pay. While some of its problems were undoubtedly of its own making, others were not. The British share of world tonnage fell from 41.6 to 34.2 percent, partly through overseas construction (especially in the USA and Japan) and partly because of the 15,000,000 tons of shipping sunk during the war, 9,000,000 tons were British, resulting in an absolute as well as a relative reduction. The important "cross-trades" were almost entirely lost during the war and not easily regained, especially in view of a market shift away from cargo liners towards tramping. The East Asian trades recovered quickly in Liverpool in the 1920s, as indeed did Blue Funnel, their principal exponents, but most others did not. For some key trades like cotton importing and coal exporting, we can see with hindsight the beginning of the end.

Seeking Reform

After the opening of Gladstone, and probably sooner, the advocates of huge facilities for large ships were effectively a spent force. But if they had damaged Liverpool's finances, and the overall balance of its physical structures by their unreasonable demands on the Board, there was no guarantee that port customers in general would behave more wisely responsibly in other ways. In 1924, the Board had the "benefit" of the publication of *Port Facilities of Great Britain: The Report of the Port Facilities Committee appointed by the Chamber of Shipping, by the Association of British Chambers of Commerce and by the Federation of British Industries*.[36] While it was comparatively complimentary about Liverpool,

---

[34] Alfred Jones' rise from having worked as a cabin boy made a particularly good story illustrative of Britain's ready acceptance of commercial innovation, hard work and self-improvement.

[35] A good deal of the history of shipping companies continued to be largely based upon the "great man" until the arrival of maritime business histories, in the 1960s, in whose production and influence F.E. Hyde and the so-called "Liverpool School" were important.

[36] *Port Facilities of Great Britain: The Report of the Port Facilities Committee appointed by the Chamber of Shipping, by the Association of British Chambers of Commerce and by the Federation of British Industries* (London, 1924).

it also exhibited the old failing among port customers of assuming that any reasonably intelligent bunch of people can run a port authority better than those who actually do it.

Some of the suggestions for modernisation, particularly in the matter of electrification, were, as we shall see below, already in hand but proceeding more slowly than the Board would have liked owing to the reluctance, yet again, of the shipowners to pay. Some, however, were foolish, and one in particular would have been a disaster had it been adopted. The Board's electrical equipment at the time was either supplied by small "local" generating plants or the Corporation's system. The Port Facilities Committee suggested that the Board should build its own central power station at precisely the time when the Corporation was considering major expansion involving power stations an order of magnitude bigger than anything the Board could possibly need even if it electrified absolutely everything, including all the graving dock pump houses, hydraulic power centres and the like.[37] If there were a worse time to build a small power station, it would be hard to find it. As is explained in more detail below, the Board rightly decided that in light of very rapid expansion in the power generation industry (the average output of generating stations rose by a factor of about twelve during the war years) it should defer the issue of power supply.

What is more important, however, is that the Committee was advocating charging through an open door. Spurred on by complaints of congestion on the quaysides, the Board had appointed a Special Committee in 1917 to investigate the possibilities of increasing throughput by installing additional mechanical handling equipment and providing power supplies for them. When the tenants of appropriated berths were offered the benefits of such things as portable electric cranes on very favourable terms, just two were willing to pay.[38] The Committee suggested improvements in the coal bunkering plant; the Board invested instead in oil bunkering facilities. Obviously the report also contained some sensible suggestions, but none had been overlooked by the Board, merely delayed by the crucial importance of finishing Gladstone and the re-construction of Alfred entrance before committing funds to other projects. As related in chapter four, such measures as cutting down the "haunches" of old passages and building out

---

[37]The Lister Drive No. 3 Power Station was one result of deliberations which began in 1922, and when it opened in 1926 its output was fifty-eight megawatts. The Pumpfields Power Station, of a mere 8.4 megawatts, had become uneconomically small by 1913 and was operated only as a seasonal standby. Still, 8.4 megawatts was much more power than the Board could use.

[38]The minutes and reports of this Committee survive as NML, MAL, MD&HB, MP/27/2.

into docks to provide wider quay margins were implemented as soon as the situation permitted.

In 1931, the Liverpool Steam Ship Owners' Association (LSSOA) published a *Report on Liverpool Port Dues on Shipping* which provided an impressive list of the woes of shipowners, which were by then both genuine and severe.[39] It illustrated their fall in income and compared it with the rise (as compared with 1913) in port dues of 72.7 percent. Little account was taken of the reductions in dues in 1922 and 1930, and even less attention was paid to the need to continue to improve the Estate. The only significant suggestion offered for reducing the port's costs was to cut back on investment. Here were people who worked in an acutely cyclical business trying to persuade the Board, with its long tradition of counter-cyclical spending, to aggravate the difficulties by behaving like them. The Board chose to make its economies in other and wiser ways.

One complaint in the LSSOA *Report* deserves a little more investigation. It complained of the rise in "General Charges," in which a major component was the General Manager's expenditure, which were shown in the *Annual Accounts* as having reached the alarming level of sixteen percent of dues income. This rise was in fact a long-term trend and was in some ways a fair reflection of increasingly expensive bureaucracy in the Board's normal work. It was a problem which the Board had addressed on several occasions, notably by the appointment of Special Committees in 1890 and 1911. It certainly was a serious matter: the General Charges for 1930 were equal to seventy-one percent of the capital works budget or forty-five percent of the maintenance budget for the whole of the Board's Estate. But since the Board had genuinely tried to cut down bureaucracy and failed, we need to find out what lay behind the continued high level of general charges. To begin with, by far the largest individual component in general charges was superannuation payments. Even had the Board wished to welsh on its former employees (and there is nothing to suggest that it did), it would never have dared do so because of the message that would have sent out about its creditworthiness. Still, in 1930 the General Manager's Department cost £38,307, a close second to the Engineer's Department's central establishment charges of £40,012 and far higher than those of any other department.

One of the reasons for this goes back to the recommendation of the 1890 Special Committee that claims against the Board involving sums of less than £5 should "be settled by the Secretary [later Secretary and General Manager] on the best terms he can" in order to try to prevent the Board's expensive Standing Committee procedure continuing to be clogged up with trivia. It is true that the golden days of trivia were past, for nothing quite of the quality of the prolonged arguments over the dues on broom handles – which were in no circumstances to

---

[39]Liverpool Steam Ship Owners' Association (LSSOA), *Report on Liverpool Port Dues on Shipping* (Liverpool, 1931).

be confused with cut lengths of square timber intended to be made into broom handles at a later stage[40] – seems to have survived the war. The majority of claims made against the Board made in the 1920s by shipowners relate to minor collisions and groundings and were typically around £200. Those made by owners of cargoes were normally for smaller amounts and commonly involved water damage resulting from roof leaks in sheds or warehouses. There was a small but growing number of claims between about £10 and £50 involving accidents to road vehicles brought about by goods falling out of slings or the failure of lifting appliances. As might be expected, the smaller the claim, the smaller the claimant was likely to be.

The sequence known as Berth Papers also contains evidence of a more sensible attitude towards the Board than at earlier dates.[41] Where before these had been cluttered with dozens of what one could only call petty whinges, in the 1920s their contents relate mainly to three issues: the construction of small offices inside or adjacent to sheds; the installation of better handling equipment, frequently electric; and the state of sheds, normally relating to roof leaks or deterioration of the floors. From this latter category, just as from the record of claims against the Board for cargo damage, it is clear that there were formidable arrears of maintenance even before the economy drive just before and during the Depression. Near the end of accounting year 1927-1928 money ran short and a number of projects were suspended, including repairs to the roof of a shed at West Float, appropriated to Lamport and Holt at an estimated cost of only £835. When it became clear that the preliminary budget for the following year would result in a deficit of around £25,200 before making any provision for the Sinking Fund, the Renewals and Depreciation Fund or the Fire and Marine Insurance Fund, the Finance Committee asked the other committees to look again at their budgets.[42] The Works Committee identified a further saving of £83,000. The following year it cut its estimate for repairs to shed floors on the Liverpool side from £20,000 to £2000. Far from being extravagant, the Board even at that stage was willing to risk building up such arrears as would constitute permanent decay

---

[40]Afficionados of trivia can follow the broom handle saga in *ibid.*, UWUP D123. The classic age of discussions about huts was also past, but occasional examples continued to appear, such as the decision of the Warehouse Committee on 12 January 1921 that it was "agreeable to the hut being repaired at a cost of £2 and afterwards being sold to the Anglo-American Oil Co. Ltd. for the sum of £10." WUP 36/2.

[41]The Berth Papers are within the UWUP sequence and consist of files, mainly correspondence, relating to the tenancies of appropriated berths. They are filed alphabetically by the tenants' names.

[42]This uncomfortable process may be traced in NML, MAL, MD&HB, WUP 202/1.

of their estate. It may be that this was the beginning of the long slide which resulted in fully half the berths in the port being in a thoroughly decrepit condition by 1970.

In fact, they were probably over-cautious, for the *Annual Accounts* for the year ended 1 July 1929 show a weak but not dangerous state of affairs. Receipts were up slightly, and it proved possible not only to add the usual £100,000 to the Sinking Fund but also to make reasonable "set-asides" to the Unappropriated Receipts Account and the Fire and Marine Insurance Fund. One mitigating factor in the apparent financial gloom was that interest rates were falling, with old bonds bearing as much as 6.5 percent being paid off and replaced with new ones at five percent or less. Ironically, it was probably exactly that combination of caution and a little good luck which made the Board's customers think they could squeeze yet more concessions.

In these dangerous circumstances, even major companies would and did occasionally waste the Board's inadequate resources and generate large amounts of correspondence, reports, discussions, time spent in Standing Committees and writing of minutes by making claims for alleged damages to vessels or cargo which varied between the small and the downright piddling. In 1922, for example, Spillers Grain Company made a claim for £1 17s 6d which was solemnly processed through the General Manager's office and then reported to the Warehouse Committee.[43] In all but the most serious cases, the Committee would express its view in general terms and then resolve that "the matter be left in the hands of the General Manager." This was an appropriate use of delegation to save Committee time, but it naturally increased the workload in the General Manager's department, which is where we started.[44] For the customers to complain of rising costs in that department was the desk-bound equivalent of the practice mentioned above of shipping lines dumping rubbish at their own appropriated berths and then complaining there was insufficient depth of water.

By the time of the LSSOA *Report* the Board was already well on the way to making significant cuts in expenditures, both on the physical structure of the port and on the cost of managing it. In addition to the measures mentioned above, in 1931 it had appointed a Special Committee, on similar lines to those of 1890

[43]The preceding passage is based on *ibid.*, WUP 9/4.

[44]The same is true of claims made by the Board for damage to its property which, particularly in the 1880s, exhibited a most prodigal use of committee time in trying to recover trifling sums; after the war the Committees were much more willing to let the General Manager and solicitor get on with it. It also adopted the practice of offering rewards of £20 to "informers" who could identify the guilty parties responsible for "hit and run" damage to buoys, etc., which would certainly have been regarded as ungentlemanly before the war.

and 1911, to "Enquire into the General Expenditure and Method of carrying out the work of the Board," but even before it met considerable savings had been identified by a "Committee of Officials" and others had already been effected. The Board had 101 fewer staff than it did in 1921, resulting in a £39,000 (11.5 percent) reduction in the salaries bill, though this was eaten into by a £26,900 rise in superannuation payments over the same period.[45] The routine repairs and maintenance budget (1931-1932) had been cut yet again, this time by £37,000. The savings identified but not yet implemented included temporary closure of Queens Graving Dock (£3300 per annum), Coburg Impounding Pumping Station (£4000 per annum), and Duke's Entrance and West Langton River Entrance (£2200 per annum). Stopping the Canada hydraulic pumping station at night would save £204 per annum. Two railway locomotives put out of commission would save £1300 per year, and two grab dredgers £1300. The really big cut, however, was to lay up *Leviathan* at an estimated saving of £40,440 per annum.

It will be seen from these examples (and they are by no means a complete list) that the Board was entirely serious about reducing expenditure; it was not just going through the motions. The Committee's first report was submitted on 21 September 1931 and included recommendations for additional cuts, including laying up the old 100-ton crane barge *Atlas* (by then scarcely used except by the Engineer's Department for bridge and gate repairs) at a saving of £1600 per annum. It even sacrificed its own "jollies" by putting *Galatea* out of commission, saving another £5000 per year. The total savings identified by both committees amounted to £151,104 per annum, to which was added a saving of £235,000 on capital account by reducing the size of Bidston Dock, then under construction. Very little that could save money escaped its notice: its final report on 23 November 1931 proposed that their own pocket books were henceforth to be bound in cloth instead of leather, saving £140.

Some of the cuts produced howls of protest from customers, notably the proposal to close Dukes Entrance.[46] Dukes was an obsolete little dock used by the barge trade, with a forty-foot half-tide entrance and (at four feet below Old Dock Sill) one of the highest sills in the port. It was, however, used by large numbers of flats and barges in the forwarding business, including Spiller, which stated that its cross-river traffic amounted to over 300,000 tons per year, and the Wolver-

---

[45]Despite this reduction in staff, the tonnage of vessels paying tonnage dues was 2,598,535 greater in 1931 than in 1921.

[46]The material in this paragraph comes from NML, MAL, MD&HB, WUP 202/1; and UWUP E47. The first suggestion to close Dukes Entrance was made in the Docks and Quays Committee on 29 July 1931. The howls of protest were mainly in the form of letters read to the same committee at its meeting of 28 August 1931. The decision to close the entrance was then rescinded.

hampton Corrugated Iron Co. Ltd. (of Ellesmere Port), which claimed that 212 of its barges had used the Dukes Entrance in 1930-1931. There was alternative access via either the Canning or Brunswick entrances but to use either of these, it was claimed, normally involved losing a tide. Dukes Entrance remained open until 1947. Closure of the West Langton Entrance also brought protests, but they were fewer and less vehement, which is perhaps surprising considering that it was a much larger and more modern entrance. Its closure was approved, and it was only re-opened when wartime congestion and bomb damage to the Gladstone/Hornby lock made its use essential. The proposed closure of Queens Graving Dock brought immediate requests for its use on specific occasions by Elder Dempster, whose appropriated berths were at Toxteth and Harrington and whose vessels were unable to use the Toxtheth-Herculaneum passage to enter one of the Herculaneum Graving docks. It seems there was a certain favouring of expenditure cuts that affected facilities used by other people.

One little incident in 1920-1921 summarises an attitude problem of shipowners towards the Board. On 5 January 1920 Coast Lines wrote to the Board asking for general repairs to the sheds it rented at South Trafalgar and North Victoria, mentioning particularly the doors, quayside lighting and floors. The Docks and Quays Committee considered the request, agreed that the work should be undertaken at the Board's expense and passed it on to the Engineer for implementation. Just over a year later, on 31 March 1921, Coast Lines was in communication again, wanting to give up its appropriation of the same sheds, but asking the Board for first refusal to have them back again if trade improved. In an uncharacteristically testy response it was told that the Board was quite happy to withdraw the appropriation but that there was no question of keeping the space free for the company: "in fact, the Board will appropriate it [to someone else] at the first opportunity."[47] Coast Lines was asking the Board, as a non-profit trust, to accept one of the normal risks which the company took in the conduct of its business in seeking a profit for its shareholders. In the shipping industry, one of the bigger risks was that over-expansion during good years would leave the shipowner with a lot of expensive excess capacity during bad times.

By 1934, early signs of recovery from the Depression were clearly visible, and the tonnage of ships using the port was already greater than in 1913, but since the Board's financial state was still weak enough to demand caution, the

[47]*Ibid.*, WUP 188/1. This WUP is one of a small number of Berth Papers files which grew too large to remain as UWUPs and were therefore bound and listed in the WUP sequence.

1931 Committee was reappointed.[48] Its report states that all the previous recommendations had been implemented and that the General Manager had identified a further annual saving of over £25,000 which had also been accepted. The Committee was becoming worried that it was cutting too much, especially on maintenance, and considered instances where it could spend capital to save revenue, including rationalising all stores and workshops into the Coburg Dock Yard.[49] On balance, it decided, the capital cost was too great in relation to the prospective saving.

Conclusion

During the period between 1870 and 1914, the Port of Liverpool suffered serious harm from the unreasonable demands of its customers, particularly the shipowners but also the owners of cargo. This situation reflected a marked improvement in the period after the Great War, particularly in the matter of numerous, trivial and often vexatious claims, requests and general correspondence, but also in less of the endless demands for expensive capital works with huge debt charges and depreciation. The lack of respect for the Board's role, evidenced by the idea that any fool can run a port, seems to have diminished little if at all, and two significant reports on port facilities devoted much of their space to urging the Board to do things it had already done.

The Board, too, seems to have effected a considerable simplification in the way it worked, helped perhaps by this intervention by Sir Richard Holt at a Meeting of Committee Chairmen:

> The Chairman referred to the trivial matters now being submitted to the Meeting of Chairmen of Committees and suggested that the Manager [sic] be asked to consider the conditions under which matters are submitted to this committee, with the view to limiting the same to cases involving substantial sums of money or matters of principle. This was agreed to.[50]

---

[48]This was reflected in the fact that there was no capital expenditure except a modest £108,426 on the continuing work on the Crosby Channel training banks. The shipping tonnage had revived, but the dues income was still lower than in 1931.

[49]Given that at the time it could borrow at three percent, this would have been an extremely good idea provided it also modernised what was already there and got rid of a few hundred tons of "old materials."

[50]NML, MAL, MD&HB, WUP 202/1, meeting of chairmen, 3 January 1929. "Substantial sums" were subsequently defined as £500 and over.

This attention to trivia, which came about partly through a desire by the principal officers to pass the buck and partly through a predilection by some members to meddle in everyday affairs, had long hamstrung the Board in its proper role of broad-brush policy making. It did not disappear, and examples of its survival will readily be found in this and other chapters, including the suggestion above that members of the Board should participate in periodical inspections of scrap heaps. All the evidence suggests, however, that there was a very real improvement.

There was, of course, one respect in which the Board could never win. If it had a revenue surplus which enabled it to make contributions to the various "funds," the customers wanted rate cuts.[51] Never mind that it was from those funds that key elements of modernisation, notably the electrification programme, were paid. If, on the other hand, it did not make those contributions, the raising of further capital would be difficult.

The Board was meant to be a two-way commercial "grapevine," gathering information as to what was needed by the customers but also disseminating, through its members, a sympathy with the Board's problems and a knowledge of what was achievable. In the latter function its failure seems near-complete: Sir Alfred Read, for example, having been a member of the Board, knew all this, but as was recounted in chapter three, this did not stop him from making repeated and sometimes unreasonable demands on the Board which, when acted upon, resulted in heavy expenditure on facilities which did not meet their own interest costs, much less allow for depreciation. There was only one possible long-term outcome if such expenditure were continued, and that was insolvency. That was where the Board was heading fairly rapidly heading in the opening years of the twentieth century, and it was only the more robust approach adopted later, coupled with a more realistic attitude by most of the customers, that left the port in a comparatively sound financial condition on the eve of World War II, albeit at the expense of much of the Estate being in a poor state of maintenance.

---

[51]Contributions were made to the Sinking Fund, Insurance Account, but above all the Unappropriated Receipts Account. Shipowners, who often held considerable, and secret, reserves clearly understood this situation and their sincerity in raising the issue is open to question.

Chapter 6
Sand, Silt and Sewage: The Work of the Board's Dredgers

From the opening of the Old Dock onwards, the port had to face the problems brought about by the very heavy silt burden carried by the water of the Mersey, a burden which it bore only when moving at more than about half a knot; if it flowed slower than that, it was deposited. The Old Dock had to be allowed to run dry periodically for "scuttling" - the manual removal of silt into barges - and dredging technology made rapid strides during the nineteenth century. The first steam dredger was put to work at Sunderland Harbour in 1797; the first steam bucket ladder dredger followed at Portsmouth in 1802 and was capable of raising about eighty tons per hour. Liverpool's first steam dredger, also of the bucket ladder type, followed in 1818 at a cost of £4294, plus £3900 for six hopper barges at £650 each. It could lift about forty-five tons per hour from a depth of about twenty feet.[1]

The bucket ladder dredger was a rather clumsy tool for dredging in or close to passages and entrances, a task which was gradually appropriated by grab dredgers, of which *Secunda*, purchased in 1859, may have been the first in Liverpool. It certainly proved durable and was not scrapped until 1928.[2] The bucket ladder dredger continued in use for general purposes but became more particularly associated with dredging rock and heavy materials in new construction works or the deepening of an existing dock. Its ability to work close to walls was improved by the introduction of the double side-ladder design, of which *Tantalus* (1900) was probably the first.

G.F. Lyster first saw a sand pump dredger (SPD) at work in the River Loire in 1876. While afterwards he conducted a small-scale experiment with one, he did not adopt the principle in earnest until he was forced to do so in 1890, when a large-scale experiment in dredging a breach in the bar began, leading

---

[1]A.W. Skempton, "A History of the Steam Dredger, 1797-1830," *Transactions of the Newcomen Society*, XLVII (1974-1976), 97-116.

[2]National Museums Liverpool (NML), Maritime Archives and Library (MAL), Mersey Docks and Harbour Board (MD&HB), Works Committee Papers, 7 February 1928. The use of "may have" and other vague expressions is necessitated by the absence of a reliable fleet list before 1897. The sources are discussed in greater detail in A. Jarvis, "The Fleet of the Mersey Docks & Harbour Board," *Northern Seas Yearbook*, forthcoming.

eventually to the construction of three large SPDs and one gigantic one. From then on, the SPDs were mainly employed at the bar, in the main approach channels and at the Pluckington Bank. In 1899 Lyster confidently explained that the problem of the bar and the channels could be solved by dredging alone, at far less expense than by the construction of training walls or other fixed works.[3] It was a confidence he would live to regret.

The Board's fleet of dredgers for each of these now fairly distinct roles expanded rapidly in the 1890s. The main new vessels built or converted were the centre-ladders *Walter Glynn* (1895) and *Lyster* (1896); the large grab hopper dredger *Miles K. Burton* (1897); the two experimental SPDs converted from steam hopper barges numbers five and seven; but above all the large SPDs *Brancker* (1893) and *G.B. Crowe* (1896). These vessels represented major expenditure from the Unappropriated Receipts Account, the balance of which fell from £845,898 in 1894 to £628,467 in 1897.[4] It is difficult to know exactly how many small grab dredgers there were at this time because no "fleet list" exists until a later date. The Insurance Account for 1893 mentions "5 dredgers," of which three must have been *Brancker* and the two converted hopper barges five and seven, implying that there were only two grab dredgers. Given that dredgers had such hard lives that they rarely managed to spend less than one-third of their time under repair (the dredging section at the Dock Yard employed about fifty millwrights, with smaller numbers of other trades), there must clearly have been others which were over ten years old.[5] A barely acceptable minimum would have been one for each district (i.e., four) plus one spare to partially cover down time. Moreover, the Board had a propensity for keeping things that still worked for a very long time. In 1902, for example, it was decided that the salvage lump *Albert* was beyond reasonable repair. It had started life as a wooden light vessel in 1840, and was survived by *Queen*, another wooden light vessel dating from 1839. In 1913 the floating grain hoist *Penmaen* was up for disposal; it was converted from a schooner built at Runcorn in 1842 A derrick flat proposed for disposal in 1962 was 122 years old, and the 200-ton floating crane *Mammoth*, built in 1917, was still in service at the time of the Board's financial collapse in 1972. With such

[3]A.G. Lyster, "Sand-Pump Dredgers," *Minutes of the Proceedings of the Institution of Civil Engineers*, CXXXVIII (1899), 394-398.

[4]As is mentioned in other contexts, it was bad customer relations to keep too much money in this extremely useful account.

[5]The principle of the Insurance Fund was that the Board paid in from general revenue a sum corresponding to depreciating the vessel's book value to zero over ten years, so that the venerable *Secunda*, for example, would not be included in the five.

traditions it is virtually certain that there were several more old grab dredgers lurking about.

As might be expected, the arrival of a large new grab dredger, which was backed up by the conversion of four steam hopper barges (Nos. 13, 14, 15 and 22) into grab dredgers, caused a rapid rise in the output of dredgings from the docks. But as mentioned in the previous chapter the capacity of the dredgers was to some extent restricted by the unwillingness of tenants to leave their berths empty long enough for dredging to be carried out. This was in part because they had not got used to the idea that berths needed dredging more often than in the past. As with so many other problems faced by the dock engineer, the increased depth of water demanded by deeper-draughted ships was the underlying cause.

Mersey water typically contained about 0.001 percent solid matter by weight, and as soon as one sees it in those terms it becomes clear that a deeper dock has a greater volume-to-surface area ratio than a shallower one and will therefore accumulate proportionately more silt. There are several aggravating factors. In a number of cases, river entrances were kept clear of silt by means of a sluicing system which released impounded water from the docks through various forms of low-level "nozzles" while the tide was falling, stirring up silt which was then swept away on the ebb. While these were successful on a small scale, none of the really large ones worked as intended, and they involved huge volumes of water. That at Canada Entrance, if all the sluices were fully opened, would release 50,000 tons of water per minute. Clearly that water had to be replaced within the system, and that happened when the entrances were opened at the next high tide. Just one minute's work at "full bore" meant that approximately half a ton of silt would settle in the dock(s) which supplied the sluicing water, in this case Langton and Brocklebank.[6] As is clear from the *Engineer's Reports,* Langton and Brocklebank had among the highest dredging costs in the entire system.

Huskisson was another expensive dock to keep dredged, and here the immediate reason was different but the underlying one the same. The endless pressure for deeper water had led, as mentioned in chapter five, to the installation of impounding pumps to maintain the level in the fully impounded docks at around high water, spring tide (HWST). They were generally successful, except for their high wages cost, but, like the sluicing systems, they introduced large amounts of additional river water into impounded docks where it would dump its silt. The quantities of water, and hence silt, were not as great, but they were scarcely trifling: the Huskisson pumps delivered a total of 1200 tons of water per minute, so that an hour's work would again deposit roughly half a ton of silt.

---

[6]This figure is calculated from information in A.G. Lyster, "The System of Sluices at the New North Works, Canada Basin, Liverpool," *Transactions of the Liverpool Engineering Society*, I (1876-1877), 13-20. Figures for dredging costs are given below.

Because they enabled half-tide entrances to be kept open longer on each tide without allowing the levels in the fully-impounded docks to decline by up to eleven feet during the days when the tides were falling from spring to neap, they were used to do just that, with the result that they pumped for about four hours a day, and sometimes four hours a night as well, on most days. This alone would introduce silt equivalent to a day's work for a grab dredger per pumping station per year. But to allow pumping for as long per tide as possible, the inlets were at a low level, resulting in a greater-than-normal proportion of silt in the water delivered. Because the pumping stations were provided where the need for depth was greatest, the effect of this additional silt was proportionately greater. In 1909 the forty-one acres of the Alexandra system, not used for sluicing water and well away from the nearest impounding pumps, cost £729 to dredge, while the twenty-one acres of Langton (sluicing water) cost £1781 and the thirty-seven acres of Huskisson (impounding) cost £1934.[7] Clearly, if dredging was costing five times as much per acre in Langton as in Alexandra, it was taking of the order of five times as long, with obvious effects on the availability of berths and hence on the overall efficiency of the dock.

The use of bucket ladder dredgers for construction purposes goes back well before the period under consideration here. They were extensively employed in excavating the Suez Canal, and when a large coffer-dam failed at Birkenhead in June 1864 it was alleged that the fault lay with G.F. Lyster, who had ordered the excavation of some spoil which the contractor had proposed to leave *in situ*, "to lessen as much as possible the subsequent dredging."[8] By the 1890s, the technique of building the retaining walls of a new dock in a huge trench and excavating the water space only enough to admit a bucket ladder dredger to finish the excavating had been adopted. After that we find fairly frequent mentions of dredgers, such as "the water was admitted on 23 April 1904, and the deepening of the dock to 22ft 6in below Old Dock Sill (ODS), about one fourth of which was completed when the water was admitted, is being completed by dredging."[9]

Bucket ladder dredgers were by this stage getting quite large and powerful (*Walter Glynn* and *Lyster* were each 190-feet long and equipped with twin compound engines of 152 nominal horsepower [NHP], *Tantalus* a little

---

[7]The figures are from NML, MAL, MD&HB, *Engineer's Report*, 1909.

[8]W.J. Doherty, "Description of Cofferdams used at Dublin, Birkenhead and Hull" *Minutes of the Proceedings of the Institution of Civil Engineers*, LI (1877-1878), 137-160. It should be noted, however, that Mr. Doherty had worked for the contractor on this job, and what Lyster was trying to do was to shift a cost from the Board to Doherty's employer.

[9]NML, MAL, MD&HB, *Engineer's Report*, 1904.

larger) and were also applied to one of dock modernisation's most hated tasks, the removal of old masonry work. The reader will recall the mention in chapter 4 of incompetent blasting at Alfred Entrance, where it is not explained what was *meant* to happen. One of the great merits of high explosives was that they could be used, in the right hands, to "shake" masonry without smashing every piece of glass for 300 yards.[10] The idea was to loosen the masonry, collapse the pieces into the water and then let the dredgers scoop them up. The more powerful ones could even excavate rock "blasted or *in situ*."[11] Imagining the strains on the lips of the buckets as they picked up substantial and irregular blocks may help us understand why there were so many old buckets in the Dock Yard scrap heap and why there were so many millwrights on the payroll. The technique was, of course, particularly valuable on projects where the old masonry was to be dumped at sea rather than used as backfill on another project.

Although the bucket ladder dredgers were chiefly used for construction, they could also be used in the river, to assist in deepening and then maintaining the deep berths at the downstream end of the Floating Stage, and to work on the fairway to Brunswick Entrance.[12]

Pluckington Bank

Just as the dredging in the main approach channels was all to go hideously wrong, so, on a less spectacular but equally damaging scale, was the new Brunswick Entrance set to be a major problem for a long time. Not for the first time, a dock engineer placed his faith in a large-scale sluicing system to protect a river entrance from a well-known likelihood of serious siltation. The Pluckington Bank was notorious for its intractability and for moving around. A.G. Lyster was old enough to know what had happened with his father's sluicing systems at Birkenhead Great Low Water Basin, Canada Entrance and the south end of the Floating Landing Stage. All had failed for the same reason: they provided too much energy close to their outlets and not enough further away.

---

[10]"Tonite," a mixture of guncotton and barium nitrate, was the explosive mainly favoured in the Dock Yard.

[11]W.A. Oglethorpe, "Modern Dredging Appliances," *Transactions of the Liverpool Engineering Society*, XXXI (1909-1910), 27-56. Although he does not specify, he is probably talking about *Vulcan* of 1904, which could raise 400 tons of stiff clay per hour in its twenty-cubic-foot buckets.

[12]In each case the great bulk of the work was done by the sand pump dredgers (SPDs), but at both these locations some of the material to be removed was too cohesive for them to lift.

Yet Lyster decided that it would be fourth time lucky and attacked the formidable bank with a weapon of proven ineffectiveness: he would dredge a deep fairway and then employ the sluicing water to maintain it. Needless to say, the sluicing system did not function as intended, the entrance needed large-scale dredging for the rest of its working life, and the cost of that dredging was a significant factor in the decision to close the entire south docks system, as distinct from just the older parts of it, in 1972. It is, however, necessary to restrict criticism of Lyster to his methods of tackling the problem. Just as his father had warned of the dangers of attempting to dredge the bar and been ordered, against his better judgement, to do it, so Lyster Jr. had warned members of the perils of Pluckington Bank. But they thought the several advantages of a deep entrance at that point outweighed the single great disadvantage, and the choice of location cannot fairly be blamed on him.

Figure 6
***Coronation* at Work in the River**

Source: National Museums Liverpool.

The new 3500-ton SPD *Coronation* was delivered in 1904 and immediately put to work on creating the fairway to Brunswick. In the next year it raised 4,582,500 tons and in 1905-1906 another 3,233,500 tons.[13] It is not always possible to tell exactly what was dredged where, but the total dredged in the river (as distinct from the channels) continued to rise rapidly, reaching 4,912,785 tons in 1909, 6,557,205 in 1910 and, after a slight reduction in 1911 no less than 7,509,300 tons in 1912. Despite this enormous effort, the new Brunswick Entrances were the scene of frequent minor groundings, of which the first seems to have been the Harrison liner *Workman* on 2 September 1908.[14] This brought about a regime of daily surveying by the Marine Surveyor's men, and the working of the entrances to a lesser depth than that of the sills. These "working depths" varied considerably and did not exhibit any identifiable trend. What is clear, however, is that it was impossible to maintain the approaches at less than two feet shallower than the sills, and four to six feet seems to have been the average.

There are also frequent reports of movement of the leading lights which guided vessels to the entrances, which indicates that not only the depth but also the alignment of the "fairway" was varying. At times the whole business verged on farce: the Pluckington Bank was partly composed of material too cohesive for the SPDs to lift, and two "eroders" - hydraulic devices "moored" on the river bed to break it up - had been in action almost from the start. On 7 July 1911 the Works Committee heard how SPD *Brancker* had grounded, and when drifting off as the tide rose, fouled both eroders and their marker buoys.[15]

Some of the measures adopted suggest desperation: from an unknown date until 21 October 1910 Lyster deployed his ultimate weapon, the 10,000-ton SPD *Leviathan,* on Pluckington Bank. This followed an experiment approved by the Docks and Quays Committee on 27 April which involved attempting to cut "a canal" through the bank "with faggots attched to a wire rope" - which in turn followed the grounding of *Willesden* on 7 April and *Irrawaddy* on 9 April. There had been an experiment employing "the sewer to the Southward of the Toxteth Lock" as an extra sluicing culvert, but perhaps the surest indication of desperation was that the old Herculaneum Entrance was cleared by the bucket ladder dredger *Tantalus* and the sluicing system there brought back into use. The intensive sluicing regime now adopted at both entrances led to a reduction in the

---

[13]This and similar figures are from the NML, MAL, MD&HB, *Reports of the Acting Conservator*, for the appropriate years.

[14]Details of groundings and working depths are to be found in *Ibid.*, Worked-up Papers (WUP) 126/1.

[15]*Ibid.*

depth of water in Herculaneum, Harrington, Toxteth and Brunswick docks, which meant that instead of ships not being able to get to the docks, they could now not stay afloat in them.

Figure 7
1:48 Scale Builder's Model of *Leviathan*

Source: National Museums Liverpool.

The remedy was at hand: the new works had made the Coburg Impounding Station redundant, and it was now to be overhauled and brought back into commission, reported to the Works Committee as having started pumping on 22 March 1911. This, of course, resulted in importing more silt into the docks themselves and was employed only to try to gain sufficient water to make the sluices work. It was presumably not an unqualified failure since its extension for a further six months was approved on 14 July 1911 and again on 2 February 1912. The extent of the problem is perhaps best summed up by the grounding on 16 June 1909 of *Vadso*, drawing a mere eighteen feet six inches. At that date the entrances were being worked to fifteen feet six inches below ODS (i.e., four feet less than the depth of the sills), but on high water neaps there should still have been about twenty-seven feet of water.

There are two possible explanations for such an appalling situation in which one of the most experienced dock engineers in the country, a man who got a great deal right, got this one so wrong. The Pluckington Bank was, as remarked above, particularly capricious. For example, between September 1910 and September 1911 the dry area of the bank at low water, spring tide (LWST) increased from 352,998 to 404,597 square yards, and "the dry Bank South of the Toxteth Lock has decreased in area and has receded to the Eastward about 270 feet, but the North Western edge has advanced into the Fairway of Brunswick Entrances about 75 feet."[16] Changes of this kind occurred constantly and had been known to do so for a long time. It is probably significant that the Acting Conservator often referred in his *Annual Report* to a channel as "improving" or "deteriorating," whereas any sort of value judgement with respect to Pluckington Bank was rigorously avoided.

Figure 8
Creation of Brunswick-Toxteth Passage, Showing the Materials Stockyard

Source: National Museums Liverpool.

---

[16]*Ibid.*, *Report of the Acting Conservator*, 1912.

There was an underlying reason for the basic error of location of the entrances: nobody actually understood exactly how water moved in such a local "system" and therefore could not predict its behaviour or that of the solids it carried around. The only reliable prediction was probably that moving water from a sluicing system would cut through sand or silt in any random direction except for that intended or desired. The building of the Dingle Training Wall, intended to increase the scour on the ebb, was another piece of guesswork, in that the scour would *cross* Lyster's fairway. Perhaps a greater misjudgement, following on after the problems at the bar (of which more below) was the assumption that the removal of a few million tons of sand per year would have the slightest effect on the malign and dynamic structure which was Pluckington Bank.

There was another problem of which neither Lyster nor the members were aware at the time. It was a qualitative one in that the dredged materials were becoming more cohesive and hence less easily raised by the SPDs, which obviously were the dredgers of choice for large volumes. The reason for this was the large and ever-growing amount of raw sewage poured into the Mersey in the confident belief that the speed of the ebb would carry it harmlessly out to sea. This view was misplaced, but what was not understood until the interwar years was that sewage solids in the water were not just offensive but co-precipitated with the fine sand and silt normally present in Mersey water, encouraging deposit and bonding them into a sticky mud. This was what had made the eroders necessary, to say nothing of the experiments with faggots and the deployment of *Tantalus* (a double side-ladder dredger of considerable power, normally used for construction work) rather than an SPD at Herculaneum Entrance.

At the Board's initiative, a high-powered committee of six chemists and an engineer was eventually appointed to look into the question. Its report ran to 351 mostly very technical pages.[17] Their prime worry was that "agglutination" was causing a permanent reduction in the tidal volume of the Upper Mersey, which risked reducing the ebb scour in the channels. For present purposes all we need note is that it seemed proven that raw sewage discharges both favoured silt deposition and made dredging it away more difficult. Needless to say, the local authorities responsible for the discharges saw matters differently and were supported by an equally scientific rival report which concluded that the harm done to the estuary was negligible.[18] It would be half a century before anything much

---

[17]*Report of Committee Appointed by The Mersey Docks and Harbour Board to Investigate the Effects of the Discharge of Crude Sewage into the River Mersey* (Liverpool, 1930).

[18]Great Britain, Department of Scientific and Industrial Research, *Water Pollution Research, Technical Paper No. 7: Estuary of the River Mersey* (London, 1938).

got done to improve the situation, during which time the Brunswick entrances continued to need constant and expensive attention.

That unforeseeable complication might provide some excuse for the misguided decision to build them there, but it does not explain why the members could trust an engineer to overcome a problem neither he nor they understood if they could not trust his judgement in trying to avoid it. Unfortunately, none of the *verbatim* minutes cover this issue, but there is some evidence that he may have played his part. The plans for the New Works were made public at a Board meeting on 14 October 1897, and the account of that meeting in the *Liverpool Courier* the following day reported the Engineer as having spoken of Pluckington Bank in the past tense: "there was a bank there seventeen years ago, but the formation was rock;" no mention was made of the possibility things might change again.[19] In the early 1880s and again in the early 1890s the bank had extended downstream sufficiently to cause ferry boats to ground when approaching the floating stage and then receded again. It may be that both Lyster and the members thought that when the situation improved it was through the sluicing and dredging which was undertaken. But it was not: it was just that the bank had moved again. In 1912, for example, following a flurry of groundings, a Notice to Pilots was published, reminding them that at LWST "the Pluckington Bank dries nearly 1000 feet out from the River Wall."[20]

The continuation of this unedifying tale for the rest of our period can be greatly compressed. Throughout the interwar years there were frequent complaints and groundings, and the Coburg pumps were put into and taken out of commission a few times more. There were several spells of closing the entrances to shipping in hours of darkness for concentrated spells of dredging, sometimes now with bucket dredgers alone, indicating the qualitative problem. At times the Engineer was instructed to deploy "every available dredger." Alteration of positions of leading lights remained frequent, reflecting such changes as "to the Southward of Brunswick Entrances, the dry [at LWST], 12 and 18 feet contours have moved to the Westward, the dry contour about 70 yards, the 12 feet contour about 120 yards and the 18 feet contour about 130 yards."[21] It will be readily apparent that a pilot or captain lining himself up for the lock was offered plenty of scope for error by such changes - which might just as easily be in the opposite direction.

---

[19]This might be thought to invalidate the partial exoneration of Lyster above, but in reality, once the decision was made, everybody had to talk up the scheme or the investors would not buy dock bonds.

[20]NML, MAL, MD&HB, WUP 126/1.

[21]*Ibid.*, *Report of the Acting Conservator*, 1924.

The costs were considerable.[22] The first full year that dredging was undertaken as a maintenance charge alone was 1910, when it cost £20,515, plus another £393 for sluicing. Costs declined for a couple of years, and then in 1913 rocketed to £31,201, plus £272 for sluicing and £4073 for running the Coburg pumps. They continued to climb rapidly during thc war years, and in the year ended July 1918 the amount lifted was 10,226,940 tons by SPDs and another 454,504 by bucket ladders. Peacetime saw a slight decline in activity, but £90,841 was spent on dredging, plus £522 for sluicing and £7687 for the pumps. In 1921, dredging cost £125,101, and the continued use of the pumps was reflected in the high price of dredging the dock itself – another £18,460. These costs may be put in context by comparing them with the total revenue of Brunswick Dock, which that year was £76,934. This, of course, is not a fair comparison, because the objective of the new entrances was to allow the use of Kings and Queens by large, deep vessels, and they earned £134,540 and £226,195, respectively. The "New Works" were far from working at a loss, but substantial sums were being wasted. Another way of looking at the figures is that in 1910, when the Board desperately wanted to get on with Gladstone, dredging costs at Brunswick equalled almost one-half of the entire capital expenditure the Board felt it could afford that year.

As time went by, financial stringencies caused reductions in the dredging effort (e.g., £37,452 in 1930), and when trade improved, dredging increased (e.g., £65,952 in 1938), but it is difficult to tell whether this was purely a matter of money or whether it coincided with any improvement or worsening in the condition of the bank. Ships continued to ground frequently, though most pilots (or ships' masters recognised as qualified to "pilot" themselves) had the good sense to arrive a little before high water and proceed cautiously, so that if they did ground they could simply sit there until the rising tide refloated them. Those ships that got stranded for a tide were usually outward bound ones that had sailed a bit late.[23] But let us stop to think what that means: the whole point of having a deep, fully-locked entrance was to permit entry and exit over more hours of the day than could be achieved with a half-tide entrance, which Brunswick demonstrably could not do.

In chapter four an attempt was made to discover any distinctive characteristics of engineering projects which went well and those which went badly, and it may seem strange that the Brunswick Entrance problem was not considered there. That is because in this case a simple and almost monocausal explanation is offered. At the time the erroneous decision to confront the

---

[22]Costs and volumes given here are from the *ibid.*, *Engineer's Reports.* for the appropriate years.

[23]As had one of the few major casualties, the Harrison liner *Architect*, which stranded and broke its back in 1933.

Pluckington Bank was made, a start had been made, as described below, on the dredging of the Mersey Bar. This, too, would soon go astray and reveal the Board's dredgers as puny toys brandished at a stern-faced and totally unyielding Mother Nature. But in 1898, when the bill went to Parliament, the work at the bar appeared well on its way to being a triumphant success. Two huge (for their time) suction dredgers had been designed and deployed, and it seemed to the Board that if they could successfully breach the bar then the Pluckington Bank was little more than a minor annoyance which could be sucked up, carried off and dumped somewhere out of the way. That is the basis of the apparent nonchalance with which the problem was approached.

Put like that, the members and their Engineer sound stupid. Yet a naive and exaggerated faith in a new, promising but unproven technology was not new in the 1890s and is with us to this day. The composite cast/wrought iron beam was a technological miracle of the 1840s – until the fall of the Dee Bridge. The pneumatic railway was going to revolutionise inland transport, and so we could go on, via nuclear-powered merchant shipping in the 1960s to the so-called "dot. com" bubble of 2000. Initial success of a new technology can be a cruel builder of false hopes, and had the decision on the Brunswick entrances been taken as little as five years later the difficulties would have been apparent, and the location chosen could fairly have been called "insane." At the time and in the context, the Board simply had too much faith in its Engineer, who in turn had too much faith in his suction dredgers. It was an unfortunate error with serious and long-lasting consequences, but an eminently understandable one. The technology that was really needed lay a quarter of a century in the future. The first attempt to model the behaviour of Pluckington Bank did not occur until 1925. In 1898, however, no one was looking for that solution because they did not recognise the problem.

The Problem of the Bar

During the period of rapid increase in the depth of ships, many harbours and dock systems suffered difficulties with rock or sandbars at their entrances. Once the use of high explosives and rock dredges became commonplace in the 1870s, a rock bar was a comparatively simple matter, because once it had been broken up, shovelled up by a bucket ladder dredger and dumped at sea it would not come back again. For example, the depth over the rock bar outside Fremantle Inner Harbour is no less now than when it was first blasted in 1896. Sandbars were, and are, a much more complex matter. The early history of attempts to provide significantly increased depth over the Liverpool Bar has been related before, rendering only a brief outline necessary here.[24]

---

[24]A. Jarvis, *The Liverpool Dock Engineers* (Stroud, 1997), 83-89.

The first serious attack on the bar began in 1890 with the use of the two steam hopper barges converted into small SPDs which were mentioned above, and within a year they had made a breach in the bar which significantly improved the depth available. This success emboldened both Lyster Jr. and the Board, and in 1893 and 1896 it purchased two large purpose-built SPDs, *Brancker* and *G.B. Crowe*, which promised and initially delivered highly beneficial results. It was this success which led the Board fatally to underestimate the magnitude of the problem of the Pluckington Bank. Lyster Sr. had warned of the dangers of upsetting the equilibrium of the bar and had been proved wrong – or so it seemed for a short while. "According to the Chairman of the Dock Board, the Bar at the mouth of the river is now practically a thing of the past," but the Board had dredged in haste and would repent at its leisure.[25]

Lyster Sr. would shortly be proved horribly and expensively right. The argument in favour of dredging a breach in the bar was that it would concentrate and accelerate the ebb tide passing through it, thus at least perpetuating and perhaps increasing the gain in usable depth. That part of the theory worked, but it unfortunately took no cognisance of the fact that the ebb could not run faster there without running faster in other places as well. The increased speed of water through the main channel within the bar began to erode the sandbanks to either side, causing shoaling and the formation of an increasingly sharp dogleg. A third, slightly larger SPD, *Coronation*, joined the fleet in 1905, followed by *Leviathan* in 1909. This vessel, by far the largest dredger in the world, could raise 10,000 tons of sand in fifty minutes and dump it in ten. But all was in vain: the channels were still deteriorating despite the constantly-growing volume dredged.

Even before *Leviathan* was delivered Lyster Jr. knew he was beaten; the bar could not be defeated by dredging power alone, and it would be necessary to construct some sort of fixed works to prevent further erosion of the sandbanks, thereby ensuring that what the dredgers removed was not immediately replaced by an equal or larger amount from somewhere else. The first phase of this, the revetting of Taylor's Bank, was decided upon in 1907, and the last load of stone was tipped in November 1910.[26] It was successful as far as it went, but it did not actually go very far, and the erosion continued beyond its ends. In 1913 the state of the channel still gave cause for anxiety, but what really puts the problem in perspective is that the year's tonnage dredged from the bar was 2,225,600 tons, as compared with 4,802,290 from the Queens Channel and 7,358,700 from the

---

[25]*Liverpool Review*, 6 January 1894.

[26]The Board took this on as an in-house job: for details see A. Jarvis, "The Mersey Docks & Harbour Board in Anglesey: Dinorben Quarry," *Maritime Wales*, XIX (1997-1998), 93-99.

Crosby Channel.[27] The Board was running at the rate of about 15,000,000 tons per year in order to stand still.

What underlay this dreadful situation was that the Board had been cajoled and bullied by its customers into attacking the bar without having any theoretical framework within which to work. There were any number of engineers who thought they understood how estuarial systems worked, but the simple fact is that they did not. Lyster Sr. had understood that the bar was a dynamic structure capable of changing its shape and size and even of moving; he also understood that the forces producing that result must be so large as to make it very foolish to tamper. I have argued in a number of places that in general Lyster Jr. was a better engineer than his father, a judgement with which I am still content, but there is no doubt that in this particular case father knew best.[28]

There was an extensive professional literature on bars and sandbanks from which emerges no hint of a consensus on even such elementary questions as where the material of which a bar was comprised came from. To some, material was removed from a bar by the ebb tide but redeposited by the flood, while to others it was the reverse or that material was deposited at the beginning of the flow, when fresh water coming down plus a certain momentum of the ebb met the beginnings of the flood, causing a pocket of still water which dropped its burden. The point for present purposes is not to reveal who was later proven right or wrong by the march of science; that information was not available to Lyster Jr. and is therefore inadmissible. The point is that he could only rely on experience and well-informed guesswork, for there was no "correct" solution at the time. It is, however, interesting to enquire why this should be so.

In 1949 J.A. Cashin presented a recent history of attempts to control what happened in the river and its approach channels.[29] His interpretation, which is implicit above, was that it was only with the successful application of modelling the estuary that it became possible to predict the effects of any particular revetment or training wall, and successful modelling was only achieved with the Second Mersey Model, constructed and developed at Manchester University in 1930 and moved to Albert Dock in 1932. There had been a previous Mersey Model, constructed within the Engineer's Department in 1923, which was presumably a response to Sir John Purser Griffith's recommendation that the "Engineer prepare a complete scheme for training these [approach] channels, with

---

[27]NML, MAL, MDHB, *Report of the Acting Conservator*, 1913.

[28]See, for example, A. Jarvis, *Liverpool Central Docks, 1799-1905. An Illustrated History* (Stroud, 1991), 142-143.,

[29]J.A. Cashin, "Engineering Works for the Improvement of the Estuary of the Mersey," *Journal of the Institution of Civil Engineers*, XXXII (1948-1949), 296-367.

the aid of experimental models, scientific enquiry and such investigations as he may consider desirable."[30] The First Mersey Model appears to have been promising but not much more, which is why the Second was built. Yet hydraulic modelling already had a history of over thirty years, and Professor Osborne Reynolds has been credited with "the first real attempt to apply scientific principles in the design of such models." in 1885.[31] L.F. Vernon-Harcourt had built a Mersey Model as early as 1888. It was, as one might expect at that date, a bit rudimentary, but it was capable of reproducing past changes in the real estuary and therefore presumably capable of predicting the effects of any artificial intervention.[32] The question is why such a cheap and apparently promising technique was not applied - or at least tried out - sooner.

Griffith was responsible for a further report in 1928, this time jointly with M.F.G. Wilson, and this mentions extensive use of the first Mersey Model.[33] The authors admitted that it was a bit rough and ready, but it "has allowed us to trace the movements of sand in the bay and the channels." This report makes damning reading, pointing out that the state of the channels was worse than that in 1890 (excepting, of course, the depth over the bar), being down to half the width and much more difficult to navigate. It goes on to state that the objective of future works should be to try to reconstruct the channels to their 1890 configuration. This is a striking endorsement of Lyster's caution over dredging the bar in the first place.[34] It also seems almost equally to be a condemnation of what had happened since then, with perhaps an implication that "scientific methods" should have been adopted sooner as a result of the 1922 report. During the 1920s, the weekly report to the Works Committee of the tonnage dredged by the SPDs generally recorded around 500,000 tons (including the outer channels and within the river).

---

[30]His report was printed and published as *Report on the Proposed Training Banks along the Crosby Channel: Port of Liverpool* (Liverpool, 1922).

[31]A.H. Gibson, "Tidal and River Models," *Journal of the Institution of Civil Engineers*, VIII (1935-1936), 699-722. Professor Gibson was Osborne Reynolds' successor at Manchester University and was heavily involved in the construction and operation of the Second Mersey Model of 1930.

[32]This model is briefly described in Vernon-Harcourt's contribution to the discussion of G.F. Lyster, "Recent Dock Extensions at Liverpool," *Minutes of the Proceedings of the Institution of Civil Engineers*, C (1889-1890), 56-57.

[33]NML, MAL, MD&HB P/2/7, "Mersey Estuary River Channels."

[34]But for an important *caveat* see note 36.

This, however, had not been Griffith's first bite at the cherry either. In 1912 he and Sir John Wolfe Barry had been called in to report on three works that A.G. Lyster had proposed, namely a bank on each side of part of the Crosby Channel and an extension of the Taylor's Bank Revetment to the west. The report was favourable, and these were constructed, though in the case of the banks work did not begin until 1923; it was to consider whether these were still the right solution that the 1922 report had been commissioned. Again the conclusion was favourable, but when the works were completed the benefits were minimal. This second endorsement of Lyster's scheme may appear connected with the fact that he and Sir John Purser Griffith "had been friends for many years and were actually boys together at Holyhead. They both learned practically the beginnings of their profession on the Holyhead Breakwater, of which Mr Lyster's father was Assistant Engineer at that time."[35] Such suspicions of "cronyism" would be ill-founded, however, because Lyster Sr. moved from Holyhead to Guernsey in 1853, when Lyster Jr. was just one year of age and Purser Griffiths was five. Still, it made a good story.

In the 1928 report it was suggested that the curve at the east end of Taylor's Bank was too sharp and that a new training wall should move the channel away from it, the extra width being achieved by by dredging away part of Askew Spit and then embanking what was left. It seems that at this stage Griffith still had little understanding of what was really happening and was proposing major works on the basis not so much of the promised scientific prediction as of experience and intuition.[36] It might unkindly be noted in passing that he celebrated his eightieth birthday that year. His explanation for the lack of improvement in the channels was that the failure to act swiftly on the earlier reports meant that during the interval the question changed, with the result that the previous answer was correct no more.

But perhaps contrary to expectations, this was not just an excuse from an old man who had fallen behind the times. On the contrary, he was the one who had first advocated modelling, and he continued to do so. It was Thomas Newell, and perhaps more particularly his principal assistant and eventual successor, Thomas Norfolk, together with Captain Mace, the Marine Surveyor, who expressed doubts about the results from the model. It seems that their doubts were

---

[35]Announcement of Lyster's death (not his obituary), *Minutes of the Proceedings of the Institution of Civil Engineers*, CCX (1919-1920), 94.

[36]Nor, it appears, had he studied old charts sufficiently: in the discussion of Cashin, "Engineering Works," R.A. Stephenson pointed out that the erosion of the corner of Taylor's Bank and formation of the reverse curve at Taylor's Spit had been going on since 1868: it was doubled in speed by the dredged cut through the bar, but not originated by it.

reasonably well founded, in that the first model's predictions of the interim behaviour of the channels as construction of the training walls proceeded were simply not borne out in real life. (It should perhaps be explained that the work was deliberately carried out slowly in order to watch the effects as it proceeded and, hopefully, tally them with model results.) Despite the occasional jibe about playing with toys, construction of the second model went ahead, and in 1932 it was installed in the central hall of the Albert Dock Traffic Office. This more sophisticated model was also "tested on the job," as the first had been, and proved much more effective.[37] It was used to design a further programme of training walls, begun in 1935 and completed in 1960.

By the end of our period the dredging required in the channels remained apparently unabated: the 1938 figure for Queen's Channel was 13.4 million tons and that for the Crosby Channel 5.9 million, compared with 10.5 million and 5.4 million, respectively, in 1910.[38] Those figures, however, show the dangers of working to arbitrary datelines, for the amount dredged from the channels was just in the process of plummeting from an aggregate peak of twenty-two million tons reached in 1934. Virtually no dredging took place during the war yet the channels remained navigable. When peace came and the backlog was attacked, the tonnage dredged peaked at seven million tons in 1947 and then fell away again; by 1980 it was just 1.6 million. Such were the fruits of the work with the second model: it came close to rendering the channels self-maintaining.

In 1936, Professor Gibson quoted Osborne Reynolds:

> I have called attention to these results because this method of experimenting seems to afford a ready means of investigating and determining beforehand the effects of any proposed estuary or harbour works: a means which, after what I have seen, I should feel it madness to neglect before entering upon any costly undertaking.[39]

---

[37]In 1941 the "Model Room" was requisitioned by the Admiralty to house an anti-aircraft training simulator, and the model was stored in the warehouses nearby: it was never recommissioned. M.K. Stammers, "A Dock Where Nothing Will Ever Be Done: Albert Dock in the 20th Century," in A. Jarvis and K. Smith (eds.), *Albert Dock: Trade and Technology* (Liverpool, 1999). Peter Lucas, a former Chief Engineer who came to Liverpool in 1968, remembers seeing it there but cannot remember when it was removed.

[38]The figures from NML, MAL, MD&HB, *Report of the Acting Conservator* for those years. 1910 has been chosen because, on the one hand, *Leviathan* was fully in service, enhancing capabilities, and, on the other, because the Taylor's Bank revetment was not yet complete.

[39]Gibson, "Tidal and River Models."

It may have taken half a century for such ideas to get accepted, and to refine the necessary modelling techniques, but they certainly worked. What is interesting about such a generally sorry tale is that criticism of the Board by its customers for a failure to act on providing greater depths of water eventually goaded the members into action which was premature, potentially threatening to the entire regime of the river and channels (and thus to the continued existence of the ports of Liverpool and Manchester) and a source of heavy and ongoing expenditure for decades. Money which was urgently needed for modernisation works was leaking away year by year to meet the insatiable demands of the customers, but such are the consequences of free competition in the ports services industry.

Chapter 7
The Provision of Specialist Cargo Facilities

Background

As a long-term policy, the old Dock Committee and later the Board preferred not to provide specialist facilities, seeking to offer only "highest common factor" quays, sheds and handling equipment which could serve equally well for almost any trade, inward or outward alike. Those wanting special facilities were encouraged to provide them for themselves. This policy, however, was bound to face demands for exceptions, of which perhaps the most obvious early example was the construction of the low quayside at Brunswick (1832) to facilitate the handling of imported timber.[1] Berths at such a quay were usable by other import trades and for export, outside of the (very marked) timber season, but only with some inconvenience, and there were, of course, no transit sheds, so that other cargoes arriving either had to be cleared from the quayside immediately or face the risk of rain damage.[2]

Another form of specialism was exhibited at Albert Dock Warehouses (completed 1846), which were designed to keep high-value and high-duty goods away from the pilferers and smugglers, to provide bonded accommodation and to offer a high degree of fire protection. They were, however, fairly versatile and could handle any form of goods which came in bags, bales, casks or crates. While intended as import warehouses only, provision was made in the structure for hoist doors on the dock side of the warehouses to allow their use for exports with a minimum of alteration.

The first completely single-purpose facility of any size was the high-level railway at Bramley-Moore, designed to load export or bunker coal and opened in 1856.[3] This carried end-tipping railway wagons from the Sandhills depot of the

---

[1]It might be argued that Clarence Dock (1830) was built specially for steamers, but all that was special about it was its separation from the other docks: its design, construction, shed provision, etc., were typical of their period.

[2]Though, of course, there were those to whom a little rain between discharge and weighing was not entirely unwelcome, as it increased the weight of the consignment.

[3]A. Jarvis, *Liverpool Central Docks 1799-1905. An Illustrated History* (Stroud, 1991), 100-116, gives further details. The first had been the small provision of "silos" (probably actually cells) in the Birkenhead Dock [General] Warehouses in 1866. By 1870 they had handled just one cargo. National Museums Liverpool (NML), Maritime Archives

Lancashire and Yorkshire Railway to the water's edge, where special-purpose hydraulic cranes lifted them in a cradle, swung them out over the ship and tipped them. It was quite successful and was considerably extended in 1884. Just when ships were getting too large to use it, the practice of coaling the largest passenger liners out in the river using barges and floating elevators gave it a new lease of life loading the barges, and its final long decline only began when the postwar slump combined with a growing use of oil fuel and intensifying foreign competition in the coal trade to slash the demand for both bunker and export coal.[4]

A much larger and riskier development occurred at Waterloo and Great Float (Birkenhead) in the late 1860s, namely blocks of very large grain warehouses, the Board's second sortie into non-mineral bulk cargoes.[5] These were the world's first grain warehouses to be equipped with an integrated, centrally-powered grain-handling system. They can scarcely be deemed successful, since their completion was long delayed by teething troubles with the machinery, which in fact never fully performed to specification, and the overspending was close to 100 percent. On completion, the Liverpool Grain Storage Company refused to accept them, and the Board was forced reluctantly to operate them itself.[6] They were then rapidly rendered obsolete by improved methods introduced elsewhere. But their failings did not result from their single-purpose nature, so while they were a grave disappointment in some respects, it was not for want of flexibility. When further capacity for bulk grain was needed, A.G. Lyster was packed off to America to find out about latest practice, and some of what he learned was eventually applied, but only indirectly.[7] When the first grain warehouse in Liverpool to use vertical silos, such as he had seen, instead

and Library (MAL), Mersey Docks and Harbour Board (MD&HB), Unbound Worked-up Papers (UWUP) C81, memorandum, 12 March 1870.

[4]This method was also used later within the north docks, as a ship could bunker at the same time as loading cargo, thereby trimming turnaround time.

[5]This project is considered in some detail in Jarvis, *Liverpool Central Docks*, chapter seven.

[6]Predecessor of the Liverpool Grain Storage and Transit Company Ltd., which was formed in 1883 and of which more below.

[7]Lyster's report is transcribed and annotated in A. Jarvis, "A.G. Lyster's *Report on the Several Methods of Dealing with Grain in the United States, and also on the Elevated Railroads of New York*, 1880," *International Journal of Maritime History*, X (1998), 155-182.

of horizontal "cells" was completed in 1886 at Alexandra, it was built not by the Board but by the Liverpool Grain Storage and Transit Company.

The early growth of the petroleum trade was very rapid, offering significant revenue - but at the risk of fire and explosion. Here, too, the Board provided special facilities, namely the "casemates" at Herculaneum, where a low cliff on the inland side allowed the excavation of sixty artificial caves for storing petroleum products in barrels. While this appears to have been undertaken largely for safety reasons, in the case of the first major bulk petroleum facility, the adjoining Parkhill Depot, it was a case of investing to kick-start what the Board correctly foresaw could become a very lucrative trade. It installed storage tanks, together with a barrelling plant and filling stands for road and rail tank vehicles. These were successful investments and opened the way to even better ones.

Another rapidly-expanding marine technology, while not rivalling the growth rate of petroleum, was that of refrigerated cargoes, particularly southern hemisphere meat and (a little later) out-of-season apples and other fruits. Here the Board was distinctly uncooperative, effectively forcing importers of frozen or chilled cargoes to provide their own facilities, as in the case of Nelson Brothers at Bramley-Moore or the Liverpool Cold Storage and Ice Company at Albert. The Board had some bad experiences with refrigeration machinery at the Foreign Animals Wharf in Birkenhead, and it is understandable that it shied away from the potential liabilities of providing another public refrigerated facility, with its high installation and running costs and its ability to destroy high-value cargo through comparatively minor malfunctions.[8]

The Foreign Animals Wharf was a different case. Here the choice lay between providing specialist facilities and losing the trade completely when the Contagious Diseases (Animals) Act of 1878 came into force. The actual operation was undertaken by tenants, with the Board providing only serviced accommodation. (including cleaning) The degree of specialisation obviously varied: some of the lairage accommodation was converted from sheds or warehouses and could be (and was) used for either purpose as the balance between different trades shifted. Slaughterhouses, chill rooms and meat stores were less versatile. Overall, despite the repeated refrigeration problems, the Foreign Animals Wharf normally made a "profit" of a few thousand pounds, sufficient to represent a small return on the investment. It was, however, a facility mostly driven by outside factors beyond the Board's control.[9]

In another specialist field, that of heavy lifts, the Board took it as read that it should make the necessary provision. At Sandon Dock, for example, it

---

[8]K. McCarron, *Meat at Woodside. The Birkenhead Livestock Trade 1878-1981* (Birkenhead, 1990), 23-31.

[9]*Ibid.*

provided a fifty-ton rolling crane (i.e., one on a short track at ninety degrees to the quayside), which was originally (1852) hand-operated, but converted to steam in 1871. There followed a huge 100-ton hydraulic crane at Langton Branch (1883), and it seems to have been the relative lack of success of this machine that caused a shift in policy towards the provision of floating cranes for heavy lift work, beginning with the 100-ton *Atlas* in 1886, which was followed by *Titan* (1896) and *Hercules* (1900). While the two newest cranes were of lesser capacity (respectively twenty-five and fifty tons) than *Atlas*, they were much "handier" and had longer reach, and in many circumstances their effective capacity was greater than that of *Atlas*, which for lifts of over thirty tons had a hoisting speed of only seven feet per minute and was not allowed to slew at all. By the interwar years virtually all heavy lifts were done by the floating cranes, the only major exception being the export of railway locomotives from Gillbrook Basin in Birkenhead, where there was a large (eighty-seven ton) fixed steam crane.[10]

It should be clear from the above that the Board's policy of not providing specialist facilities was breached in many places, though it often seems that members would have preferred to retain it. Changes which were already well in train in 1905 made that impossible in several of the key trades of the port, so it had become and would remain a matter of treating individual issues on their merits. During the period covered here, the importance of specialist berths as compared with traditional general cargo berths steadily rose, a process we may see as heading towards the present-day situation where general break-bulk cargo represents only about one percent of the trade of the port.

Corn

So little has been written about the corn trade in Liverpool that this section will have to break the timeframe of most of the rest of the book in order to explain the antecedents of the considerable developments which took place in the early part of the last century. This is curious, since in volume terms corn far outstripped more prominent imports such as tobacco or timber. By the end of our period cotton, on which large amounts have been written, amounted to less than one quarter of the tonnage of corn, and the trade's growth to this position had deep roots which have remained unexplored apart from my own brief and rather narrow account of Waterloo Corn Warehouses. Even Graeme Milne's excellent

[10]This crane had a number of different safe working loads published over the years. Eighty-seven tons is the figure from the 1920 *Yearbook*.

survey of Liverpool merchants in an earlier period has no index entry for the corn trade.[11]

The reduction in the area of British land under arable farming combined with a reduced yield per acre and a continuing rise in population to make the corn importing trade a profitable place to be for most of the late nineteenth century. Between 1870 and 1914 wheat imports rose a little over three-fold, and when the postwar recession struck in 1922, grain imports suffered with everything else, but still remained nearly twenty-five percent higher by total value than in 1913.[12] Furthermore, the emergence of the Liverpool futures market sometimes (depending on the state of the market) provided an additional incentive to store corn. One might expect that as a result the Board would have been enthusiastic about providing new, improved and potentially profitable facilities for the storage of grain. Its general warehouses after all were quite profitable at face value and not even all that bad when allowance is made for obsolescence.[13] That, however, is a misinterpretation of its reasoning when making exceptions to the general policy of "build it yourself." At Waterloo, as at Herculaneum and Parkhill, the purpose was not to profit marginally from a flourishing trade but to encourage a completely new one.

Still, the Waterloo and Birkenhead warehouses really had gone badly wrong and left the Board with a long-standing problem. While they earned an operating surplus, £21,562 in 1884 for example,[14] their interest cost at the typical bond interest rate of that year, in round figures £38,750, turned that into a significant loss before allowing for the fact that they were already somewhat obsolescent. In some years the Birkenhead Corn Warehouses, practically alone among all the Board's warehouses, actually suffered an operating deficit.[15] Waterloo, despite its supposedly highly-mechanised operation, actually spent minimally *more* (in 1888, for example, seventy-eight percent as against seventy-

---

[11]G.J. Milne, *Trade and Traders in Mid-Victorian Liverpool: Mercantile Business and the Making of a World Port* (Liverpool, 2000). The same is true of another work noted for its study of networks, Gordon H. Boyce's *Information, Mediation and Institutional Development: The Rise of Large-Scale Enterprise in British Shipping, 1870-1919* (Manchester, 1995).

[12]It must be remembered, however, that the price of wheat had increased by fifty-one percent.

[13]A. Jarvis, *The Liverpool Dock Engineers* (Stroud, 1997), 61.

[14]These and similar figures are extracted or calculated from NML, MAL, MD&HB, *Annual Accounts*, various years.

[15]For example, for the three consecutive years 1898-1900; *ibid.*, 1898-1900.

seven percent) of its receipts in wages than the average for all the warehouses. Gradual but accelerating casualisation of what had started out as a workforce of relatively well-paid "established servants" had soon been introduced but had done little more than prevent even heavier costs.[16] This would not matter if the combination of expensive mechanisation and heavy labour costs was reflected in an atypically high dues revenue for the dock itself, justifying the warehouses as a loss-leading benefit to the customers, and this was so, to some degree, but not to a sufficient one. The total dues income of Waterloo Dock in the 1880s varied in the region of £30,000 per year, and that sum includes not only shipping and goods both inwards and outwards but also the trade of West Waterloo. An allowance for depreciation on the cost of the warehouses of just one percent would show a serious loss on the business of the whole dock and warehouses together.

The disputes over teething troubles with the machinery had got quite rancourous and were well within the memory of members still serving on the Board when the time came to make additions to the storage available to the corn trade.[17] It seems surprising that when the corn trade first started agitating for accommodation at the "New North Docks" under construction in 1879 the Board seemed quite willing to give it serious thought, as young Lyster's voyage, mentioned above, testifies.[18] Proposals and counter-proposals drifted to and fro with the over-caution which sometimes characterised the Board's dealings. In 1881 there was a discussion of warehouses or silos "which the Corn Trade propose to establish in a position contiguous to the New North Docks."[19] By the time the Liverpool Grain Storage and Transit Company Ltd. came into being in 1883 there were still no corn warehouses at the north end, but the pace of action was about to change.

By April the Company had a draft agreement with the Board whereby it would build a substantial block of brick "honeycomb" silos on the inland side

---

[16]The same happened at the Board's general warehouses.

[17]This would have included Alfred Holt, Thomas Harrison, Horace Turner, John Brancker and S. Sandbach Parker.

[18]It got dangerously close in 1879 when it agreed upon a site for the proposed warehouses but concluded "that the plans and extent of expenditure must be reserved for future consideration," NML, MAL, MD&HB, Docks and Quays Committee, minutes, 30 April 1879.

[19]*Ibid.*, Worked-up Papers (WUP) 205/1, memorandum of a meeting of the Works and Docks and Quays Committees, with deputations from the Corn Trade Association and the Liverpool Steam Ship Owners' Association (LSSOA), 7 November 1881.

of the Dock Road, served by elevators at the east ends of both No. 1 and No. 2 branches of Alexandra Dock, with a 200-foot appropriated berth at each. The reason such modest berth provision was acceptable is considered below. The Board would construct a "band alley" - i.e., a small tunnel housing conveyor belts - for their use, passing under the east quay and the road, and the Company would pay for its use. This must have been seeming precipitate to the Board, for the Works Committee approved the draft on 13 April, only for the Board variously to refer it back or postpone its consideration every Thursday until 5 July, when it was finally confirmed. On 17 February 1886 it was reported to the Docks and Quays Committee that SS *Kepler* had landed the first grain at Alexandra, though for reasons unexplained it had ended up in the adjacent transit shed (which contained the elevator) instead of in the silos.[20]

Next, however, the Company needed a base in the south docks, and on 23 September 1898 the Works Committee considered whether it ought to be a condition of any lease of land to the Company that it should either rent Waterloo or guarantee the Board's present revenue from it. This was a most curious worry to afflict the members because in the case of Alexandra it was a condition of the agreement that the Company's charges for services should in no case be *more* than the Board's charges at Waterloo. At the Company's chosen site at Coburg it seems it was more concerned about being undercut. In the event, the Board was more generous than its Committee and agreed to the lease of land at the east and south-east sides of Coburg Dock.

There were three reasons for this, none of which had anything to do with the kindness of its heart. The first was that Waterloo was obviously doomed in the long term anyway; its sill was at seven feet eight inches below Old Dock Sill (ODS), giving about twenty-three feet at mean high water.[21] The second was that a surprisingly large amount of corn still arrived in bags, and these were all-too-effectively handled at the two-story sheds at Toxteth and Harrington, which were designed for more profitable general cargo. Finally, it is reasonable to assume that the Board's corporate ear had been close enough to the ground to recognise the possibility of competition from the Port of Manchester, competition which materialised shortly afterwards with the building of a grain warehouse in 1899

---

[20]For a full description of the building and its machinery, see W. Shapton, "Description of the Warehouse and Machinery for the Storage and Transit of Grain at the Alexandra Dock, Liverpool," *Proceedings of the Institution of Mechanical Engineers* (1891), 372-385.

[21]In the event they survived in use unchanged until 1934, when the east block was converted to store oil seeds. More surprisingly, the *Engineer's Report* for 1938 records enlargement of the bottom bands of the west block, and the old (steam) engine house, having been threatened many times with modernisation, carried on pumping until 1940, when it was bombed.

and three large flour mills at Ellesmere Port, all of which began production between 1905 and 1910.[22]

Figure 9
Coburg Granary

Source: National Museums Liverpool, Ref. No. N79-476.

Construction at Coburg did not begin immediately, but the contractors (Pattinson and Sons) were on site by the end of 1902. There were the usual minor arguments over such things as the alignment of railway connections, but work generally proceeded fairly well, and the first grain was discharged into the brick silos in October 1906. The building had a total capacity of 62,000 tons of wheat and was equipped with four "marine legs" (long-reach bucket elevators) capable of discharging 200 tons per hour each. There were also chutes for loading bulk or bags into road or rail vehicles or into barges at the rate of 1000 tons per day in each case.

---

[22]The third of them, Frost's Mill, produced eighty bags per hour.

The reader who has looked at the plan of the docks in this book will already be puzzling over the Company's choice of location. Coburg Dock was fairly long and narrow, with the passages to Queens and Brunswick Docks close to its east end. The length of the east quay was only 260 feet, as compared with 1385 feet at Waterloo, whose warehouses had a capacity of only 30,000 tons, less than half that of Coburg. Large vessels could not berth at the Coburg silos at all, and that first cargo mentioned above had actually arrived in barges, not in a ship. The apparent folly of building a corn silo with a berth which could not take vessels from the most important branch of the trade, that on the North Atlantic, in fact simply reflects a fact of corn trade life at the time, namely that surprisingly little imported corn was landed by mooring an ocean-going ship alongside a corn silo and discharging its cargo directly into the cells. The idea of discharging first into barges and then, after a short "voyage" from one berth to another, into a silo, simply does not fit with our imagined view of the trade, yet it was widespread for two reasons.[23]

The highly efficient transhipment, loading and discharge which young Lyster had admired over twenty years earlier depended on the willingness of consignees to purchase grain on the basis only of a grading rather than by examination of a sample of grain of a particular origin. Under the American system, once it had been graded, grain from different sources and for different consignees could be mixed indiscriminately in handling; only the grade mattered.[24] Some purchasers still did not like this and they preferred to buy by sample and have their grain kept separate from everyone else's. This meant, of course, that a "bulk grain cargo" could not simply be continuously discharged into a silo until the ship was empty; the grain had to come out as distinct consignments. In these circumstances the use of barges was not only less inefficient than might be expected but could be positively advantageous. Nor was the situation peculiar to Liverpool: at Millwall Docks, for example, imports were "almost entirely in small parcels, which have to be kept apart."[25]

The second reason was that although by this time a high proportion of grain was carried in bulk, comparatively little was carried as whole cargoes

---

[23]And long remained so: even when the South Docks had been closed and the Brunswick Entrance left permanently open, a little grain traffic was still worked tidally by barges until 1975.

[24]This was also, of course, a prerequisite of the development of the futures market.

[25]M. Mowat, "Some Recent Grain-handling and Storing Appliances at the Millwall Docks," *Minutes of the Proceedings of the Institution of Civil Engineers*, CLXXVII (1908-1909), 59-92.

because of the severe safety problems if it shifted at sea. Mounting concern led to growing legislative control; after a rather feeble effort in the "Plimsoll Act" of 1876, the Safe Carriage of Grain Cargoes Act (1880, consolidated in the great Merchant Shipping Act of 1894) ordained that a maximum of seventy-five percent of thc total cargo of a general cargo vessel might be comprised of bulk grain if the remaining twenty-five percent were of bagged grain, or fifty percent if the remainder was general merchandise. These restrictions did not apply to purpose-built vessels of cellular construction designed to prevent grain shifting, and it was customary for it to be at the bottom.[26] On arrival in Liverpool, therefore, the ship could not conveniently go to a specialist grain berth until it had discharged whatever was on top of the corn. Once it was at a general discharging berth it might as well stay there, and in 1928 we find that:

> With the exception of vessels carrying full cargoes or large parcels of grain, all ships discharge overside into barges. Bulk grain is transferred from the vessel's hold into barges alongside by floating elevators.... barges are brought alongside grain warehouses where their contents are elevated to storage.[27]

This is why Liverpool Grain Transit had asked in 1894 for the appropriation of a general cargo berth, preferably in Hornby, "for transhipment of grain to barges."[28]

A Digression into Barge Traffic

In the late 1920s Edward Cotter listed fifty-eight floating grain elevators available for hire in the port. The Board itself had eight steam grain barges, each capable of carrying hundreds of tons, and there were many other providers of barges, steam or dumb. There were still some Mersey sailing flats in the trade. As Magnus Mowat had pointed out, it was possible for these appliances to operate

---

[26]For a more detailed explanation of a complex topic, see D.M. Williams, "State Regulation of Merchant Shipping 1839-1914: The Bulk Carrying Trades," in S. Palmer and G. Williams (eds.), *Charted and Uncharted Waters* (London, 1982). There had been a measure of non-legislative control much earlier: in 1862 Lloyd's Agents in both Montréal and New York issued notices forbidding cargoes consisting of more than two-thirds bulk grain or, at New York and in the case of larger vessels, one-half. Reproduced in R.W, Stevens, *On the Stowage of Ships* (London, 1869).

[27]E.P. Cotter, *The Port of Liverpool, Including Birkenhead and Garston* (Washington, DC, 1929), 130.

[28]NML, MAL, MD&HB, Works Committee, minutes, 30 November 1894.

over the side while other cargo-handling operations were proceeding on deck.[29] Cotter shows the process taken a stage further: on page 131 of his book is a picture of a passenger vessel attended by two grain elevators, one coal elevator and their attendant barges - and two quayside cranes working one of the after hatches. The amount of grain carried by passenger vessels - effectively as ballast - seems to have been much larger than is generally realised: Harvey Woodward, corn broker, estimated it at fifty percent of total imports when giving evidence against the 1885 Manchester Ship Canal Bill, while Dyos and Aldcroft gave a figure for 1924 of no less than eighty percent and in 1926 "the tendency is for more and more grain to be carried as parcels in liners."[30] To these vessels more than any other type, speed of turnaround was crucial, which not only meant that they would not be visiting silo berths but also that overside discharge to barges, allowing them to do two or more things at once, was particularly desirable.[31]

The barge traffic within the docks and between Liverpool and Birkenhead is extremely difficult to quantify. Only the biggest and best barges were registered at Lloyds - and some of them were decidedly large. The Liverpool Grain Storage and Transit Co. had eight vessels on register in 1908 ranging between 161 and 207 register tons, while of Liverpool Lighterage's thirteen vessels, one was a monster of 257 register tons which could load something of the order of 1000 tons of wheat. Yet it is clear from occasional claims for water damage which came to the Board's attention that there were many others of which Lloyds would have taken a dim view even if the owners had wished to have them "classed."[32] The Board itself retained some very old vessels for use within the

---

[29]Cotter, *Port of Liverpool*; and M. Mowat, "Some Recent Grain-handling and Storing Appliances at the Millwall Docks," *Minutes of the Proceedings of the Institution of Civil Engineers*, CLXXVII (1908-1909), 59-92.

[30]Great Britain, Parliament, House of Commons, *Manchester Ship Canal Bill, Session 1885: Minutes of Evidence and Proceedings (Opponents), 6 July to 3 August 1885*, qq. 14,198-14,206: H.J. Dyos and D.H. Aldcroft, *British Transport. An Economic Survey from the Seventeenth Century to the Twentieth* (Leicester, 1969), 260; and *Shipping World*, 13 October 1926. Neither Woodward's evidence nor Dyos and Aldcroft's statement is entirely unambiguous about the distinction between passenger and cargo liners, but *Shipping World* is.

[31]There was also a financial consideration: master porterage charges were much lower for overside discharge.

[32]Some of these appear in NML, MAL, MD&HB, WUP 196.

docks which were neither classed nor insured.[33] It is also clear that, although they may now be hard to identify or enumerate, barges must have been extremely numerous. The Board, for example, used to pass on some of its lighterage work to a firm called Bennett's Haulage, Warehousing and Wharfage Co. Ltd., which had nothing on register. It is also scarcely likely it would trouble to deal with a firm which had only one or two boats - the volume of business was far too large. Yet it also contracted work to T.H. Burton and Son, which had one vessel, *Helena,* on register. The efficient operation of one floating elevator required several barges, the exact number depending on the elevator's rate of delivery, the distance they had to go and the speed of the discharging machinery at their destination.

A really thorough analysis of the dock registers might enable some sort of approximation of numbers and movements, but this has not been attempted. The reason is that a brief exercise in sampling showed virtually no rise in grain arrivals at Coburg when the granary - the biggest in the world at the time - was opened, suggesting that more extensive study would be, at best, inconclusive. The dock registers are a huge and compendious source, but what they lack is the contemporary instructions issued as to how they were to be filled in. The only tentative explanation I can offer to the obvious anomaly at Coburg is that because dock dues were payable at the dock in which the ship discharged its grain overside into the barge, it was deemed unnecessary to record the cargo a second time. No charge was payable to the Board for the barges' movements unless they used a river entrance, in which case they were recorded. (This was presumably why Liverpool Grain Storage wanted facilities at each end of the system; as of 1900, when Georges Dock was closed and infilled, there was no longer any toll-free "inland route") It is clear that movements were very numerous and probably complex as well: Spiller claimed that it alone moved 300,000 tons per year in and out of Dukes Dock, whose grain warehouse was both old-fashioned and, with a capacity of 11,000 tons, relatively small.[34]

It may be worth further emphasising the large scale of barge work by a short digression. In 1901 the Board opened its new tobacco warehouse at Stanley Dock, the largest such warehouse in the world. It has vaults below the quayside and twelve floors above, a total floor area of about twenty-six acres and a capacity of 180,000 hogsheads. It is built almost to the water's edge and has continuous sliding doors for a length of 625 yards. But it handled almost no ships: to begin with, the entrance to Stanley Dock had not been enlarged since it was

---

[33]See chapter six and A. Jarvis, "The Fleet of the Mersey Docks & Harbour Board," *Northern Seas Yearbook*, forthcoming.

[34]This is in a complaint about the proposed closure of Dukes' Entrance; NML, MAL, MD&HB, Docks and Quays Committee, minutes, 18 August 1931.

opened in 1848; it was only fifty-one-feet wide with a depth of about twenty-one feet on the sill. More to the point, the Board had agreed not to open the double-decker swing bridge which carried the Overhead Railway during the hours the trains ran – but it had two lifting bascules in its lower deck to permit the passage of boats and barges.[35] The Board's Revenue Statements show that this massive warehouse investment brought about an increase in the total revenue at Stanley Dock only from £2368 in 1900 to £2825 in 1903.[36] The Revenue Statements record dues payments made on ships and their cargoes, so it is clear that little or none of the contents of the tobacco warehouse arrived by ship. The quay and the sliding doors were for the use of barges, and since by 1915 it was necessary to build an annex to the warehouse we may fairly assume that there were large numbers of them. This conjecture is confirmed by a rise in the operating surplus shown for Stanley warehouses from £14,754 in 1900 to £46,721 in 1903. The new warehouse was flourishing, even if shipping in the dock was not.

If such was the case, it should be that at Coburg, Liverpool Grain Storage's investment in the new silos would bring the Board very little yield in increased dues. In 1904, dues at Coburg were £9786, and by 1908 they had *fallen* to £8608, indicating a positive displacement of ships by barges. The fall in dues income is particularly compelling evidence since the Brunswick entrance had re-opened in the interim after five years of being closed for reconstruction, so in any normal circumstances the income at Coburg ought to have risen dramatically. As we shall see, the apparently inefficient way of handling grain which caused this anomaly still had a long time to run.

While it may not be possible to quantify barge traffic or enumerate barges with much accuracy, it is possible to give some figures indicative of the scale of the traffic. Barges may not be in *Lloyds' Register* but they are in the Liverpool shipping registers, and Jim Macdougal has extracted lists of them.[37] Of some 350 barges newly built and/or registered between 1915 and 1960, the great majority belonged to firms working in and immediately around Liverpool and Birkenhead Docks.[38] Among seventy-five operators appear such names as Kelly's Barges and Colin Jones, both of Liverpool, whose existence is not revealed in other sources. We may find some barges specific to the corn trade, including six

---

[35]This agreement was reached in 1891, before the bridge was constructed.

[36]The 1900 figure was 0.2 percent of the revenue of the whole dock system, and the increase lifted Stanley from being fourth to fifth-from-bottom earner.

[37]NML, MAL, J. Macdougall, "Liverpool Shipping Records" (unpublished mss.).

[38]No distinction is drawn here between flats and barges.

for Liverpool Grain Storage and one for Paul Brothers, but by far the largest buyers were non-specific, in the earlier years Abel, later on Liverpool Lighterage.

Even the registers, however, will not serve to give us a true enumeration of all the barges in service at one time. Take the example of *Mossdale*, registered to Richard Abel in 1920. *Mossdale* survives at the Boat Museum in Ellesmere Port, and was already old in 1920, having been bought by the Shropshire Union Canal Carrying Co. in 1867 - and it was not new then, either.[39] Its date of build is still unknown, and since we cannot know how many similar veterans carried on to very advanced ages, we cannot set an accurate rate of attrition against new construction or registration.[40] But it was most unusual for a barge to last less than twenty years, making it clear that there were several thousand barges either based in or regularly visiting the docks at the time in question. Even vessels like *Mossdale* might be found carrying grain on the Bridgewater Canal which they had collected not from Ellesmere Port but overside from a granary in Liverpool or Birkenhead or from the Bridgewater Company's own warehouse at Dukes Dock.[41] It still sounds an enormous number, but browsing the MD&HB photographic collection yields many views in which a dozen or more barges may be seen clustered around ships, and a small number depict the unglamourous places in obsolete docks where many more of them went to tie up when not working.

In this picture of guesswork and gloom there is one shaft of light. When the first Mersey Road Tunnel was proposed, the Board obviously scrutinised it very carefully with a view to protecting its own interests, paying particular attention to any loss of income it might cause. In the draft proof of evidence from Richard Holt in favour of the 1925 bill there is some tabulated information on barge traffic, and while a decline is evident over the years 1904, 1914 and 1924 for which figures are given, the trade is revealed as very large.[42] The total number of "flat-and-barge" movements through river entrances for 1904 is given as 158,294 on the Liverpool side and 58,399 at Birkenhead, 216,693 in all, and the corresponding figures for 1914 are 160,283, 49,951 and 209,434. By 1924 these had declined to 135,593, 33,456 and 168,049. Within these figures there is

[39]M.K. Stammers, *Mersey Flats and Flatmen* (Liverpool, 1993), 192.

[40]In many cases the registers *do* give the original dates of build of some very old vessels - but when the only one which has survived in a museum happens to have dropped out, it rather shakes one's faith in the system.

[41]This warehouse was only acquired by the Board in 1900.

[42]The proof of evidence, together with much else, is in NML, MAL, MD&HB, Legal B74.

no indication of what proportion of the cargoes carried comprised grain, although the fact that in 1925, seventy-five percent of the flats and barges leaving Brunswick Entrance were bound "upriver" may indicate imported grain headed for the mills at Ellesmere Port. The proportions overall (i.e., for all entrances) were twenty-five percent moving from one dock to another on the Liverpool side, thirty-four percent cross-river and a surprising forty-one percent upriver, though some of that would be raw materials bound for the Lever empire at Port Sunlight and Bromborough. Bromborough Dock was not yet open so Lever Brother's palm oil and kernels arrived at the Elder Dempster berths at Toxteth and Harrington and were forwarded by barge. Its whale oil mostly arrived in bulk in Birkenhead and was forwarded about fifty-fifty by tank barge and railway tanker.[43]

These figures, it should be repeated, represent only those flats and barges which used river entrances; those which moved around within the Birkenhead system or within Liverpool docks between Gladstone and Princes and between Canning and Herculaneum passed completely unrecorded. But when we think back to the proportions of imported grain carried on passenger liners that berthed elsewhere than at Alexandra, the unrecorded barge traffic between their berths and the Alexandra (or possibly Waterloo) granaries was clearly considerable and may reasonably be assumed to have the same number of digits as the recorded traffic. Flat and barge traffic was, in short, vastly larger than has previously been realised.

Back to the Corn Trade

Back at Alexandra, the No. 2 granary had been built in 1895, with an elevator at the head of No. 3 Branch that, with the increased capacity at Coburg, sufficed the trade until the 1920s, not least because there had been a move towards constructing mills alongside grain berths, of only modest size on the Liverpool side but on a very large scale in Birkenhead.[44] In 1900 there were just two millers alongside dock berths there, namely Buchanan and Vernon. The latter was a considerable

---

[43]Great Britain, Parliament, House of Commons, Select Committee on Private Bills, Lever Brothers (Wharves and Railway) Bill, 1920, evidence of C.C. Knowles, Managing Director of Lever Brothers Ltd., q. 1, *et seq*. There is a copy of this testimony in NML, MAL, MD&HB, Legal C62/1.

[44]This was a fairly general tendency at the time. Mills were built at comparatively minor ports such as Barry and Silloth; conversely, Southampton was the only significant port not to have any large flour mills.

undertaking: the machinery was driven by a 1200-horsepower steam engine.[45] Paul Brothers added two mills in 1903 and 1910, but such other mills as there were in Birkenhead were away from the docks and relatively small. Soon after the war, Rank and Spiller and Baker arrived, and Vernon built a completely new mill and silos, completed in 1922.[46] There were numerous lesser additions – a new silo here or there in the same period. Within twenty years the Seacombe Dock Road came to resemble a canyon, flanked for over half a mile by dour industrial buildings, all of them over eighty-feet high. In 1904 the capacity of the mills totalled 265 sacks per hour, rising to 392 sacks in 1914 and 438 in 1924.[47] Seen from across the river, they dominated the landscape of Birkenhead and Wallasey, especially before they had a rival in the shape of the Mersey Tunnel ventilator at Morpeth Dock. The old corn warehouses were extensively modernised, with new 100-ton- per-hour elevators (double the rate of the old ones), but more radically one section of them was completely gutted in order that Vernon could build eighty-five ferro-concrete silos inside to replace the "flats."[48]

These developments did not particularly benefit those of the corn traders who did not have significant interests in milling and found the futures market getting squeezed by millers who bought "afloat," especially during the shortages of the latter part of the Great War.[49] Better times returned for them in the late 1920s and early 1930s, when highly unstable trading conditions for corn made it a business in which the non-specialist dabbled at his peril.[50] In 1929 – not a year

---

[45]There is a short description in the account of the Institution of Mechanical Engineers' visit to Liverpool in *Proceedings of the Institution of Mechanical Engineers*, (1909), 867-68; the *Engineer's Report* for 1900 gives its output as 350 tons per twenty-four hours.

[46]Despite the fact that they were built by contractors for the millers, and the Board's only interest was in renting the land to them, these completions were mentioned in the *Engineer's Reports* for 1920 and 1922, respectively.

[47]NML, MAL, MD&HB, Legal B74, proof of evidence of Richard Holt on Mersey Tunnel Bill, 1925.

[48]*Ibid.*, *Engineer's Reports*, 1897 and 1900.

[49]In general, millers were also traders, but the majority of traders were not millers. For an idea of the relationships see G.J.S. Broomhall and J.H. Hubback, *Corn Trade Memories, Recent and Remote* (Liverpool, 1930).

[50]This is obviously a gross simplification: for more detail see Anon., *The Liverpool Corn Trade Association, 1863-1963* (Liverpool, 1963). Browsing the "Miscellaneous" section of *Corn Trade News* can also be informative.

one normally associates with people making large investments – Liverpool Grain Storage installed fixed pneumatic elevators of much greater capacity than the previous ones at Alexandra and built two very large transit silos on the river side of the Dock Road. The Company had already benefited from comparative slackness in other trades in 1926; seeking a berth to use for direct discharge from ship to silo, it had succeeded in persuading the Board to allow a conveyor inside the east end of the south shed at Alexandra No. 2. In times of better trade, such an idea would naturally have been dismissed as "not to be entertained."

But that very instability provided the expert in corn futures who fancied himself a gambler with opportunities to buy cheap, store and then sell dear, since the demand for storage kept on rising. In 1930 we find general warehouses at Alfred and Wallasey Docks being temporarily converted for bulk grain storage.[51] The Chairman's introduction to the *Annual Accounts* for 1931 states that "they do not make very cheerful reading," but a little bit of cheer came from bulging warehouses whose operating surpluses had risen rapidly, in some cases doubling in the last three years.

There cannot have been many firms which declared an all-time record profit in 1932, but Liverpool Grain Storage was one of them, and it was investing heavily in new facilities at the time. The development of specialist vessels for bulk grain meant that, at last, the port economics of barging almost everything were able to change, and berths where ships could get alongside quays with elevators direct to silos were becoming more desirable. At Coburg there was no really convenient quayside at which to apply the same rather impromptu solution as at Alexandra; the nearest suitable berths were at the north end of the east side of Brunswick, and they were appropriated to R.P. Houston and Co., with T. and J. Harrison to the south. Once again, in normal circumstances there would have been little question of trying to persuade either of these two famous firms, engaged in exactly the type of high-value general merchandise services which was particularly favoured by the Board as its "bread and butter," to move in favour of the relatively low-value bulk grain trade. But grain was by far the largest import in tonnage terms and it seemed to be almost completely recession-proof, and Houston and Harrison were persuaded to move, making available the site of Liverpool Grain Storage's boldest development yet.[52] It seems possible that at this stage all the Company envisaged was a large elevator and a conveyor to the nearby Coburg silo, but the following month it was definitely considering building another silo, and by April the Board had agreed to lay in an electrical supply for

---

[51]NML, MAL, MD&HB, *Engineer's Report*, 1930.

[52]*Ibid.*, Docks and Quays Committee, minutes, 23 December 1931.

this with a maximum capacity of 900 kilowatts, which seems a lot of power if all it needed was an extra elevator or two and a conveyor to the existing facilities.[53]

When the request for a site for further silo accommodation eventually came, it referred to either East Alexandra No. 1 or North East Brunswick, and on 23 January 1935 thc Docks and Quays Committee expressed itself fairly strongly against the latter, fairly favourably towards the former and very favourably to the recently-opened and very little-used Bidston Dock. Within a month, the Company submitted its plans for Brunswick: it was to be a huge building with 140 bins, each ten feet six inches by eight feet six inches and 116 feet high (the elevator and machinery tower was 235 feet high). It may be that the building could have been built on either site, but what is clear is that it could not have been designed from scratch, and the Company's decision to adopt it taken, inside that month, and the electricity supply was there waiting at Brunswick. The Harrison Line was again approached and proved quite willing to give up some berth space for the construction. Trade was still poor and it was probably quite happy to shed some of its rental expenses; certainly it did not ask for alternative accommodation. The Board was easily enough convinced; while some weeks passed in haggling over the rent, a draft agreement was approved by the Works Committee on 12 April. By November planning was nearing completion when William Thornton and Sons Ltd., the contractor, approached the Board with regard to certain roadways close to the site being temporarily closed.[54]

The reason for this rather bold request, which was granted, was that an unusual constructional technique was being employed which required the uninterrupted delivery of 600 tons of materials to the site every day from commencement to completion of each of the twin silo sections of the building. The shuttering was erected complete on the ground level of the silos, the reinforcing bars inserted and concrete pouring started. As it proceeded the shuttering was gradually jacked up, and the reinforcing bars extended so that the building was, in effect, "extruded" at the rate of six feet per day. Two shifts of 170 men worked continuously. The first section, which was to hold 60,000 tons of grain, employed 535 tons of steel, 11,000 tons of crushed granite and 2150 tons of cement.[55] On 18 March 1937 the Traffic Committee was informed that SS *Iddesleigh* had discharged the first 9100 tons of grain. The 1932 improvements at Alexandra had been hailed by the Liverpool Corn Trade Association as making

---

[53]*Ibid.*, Works Committee, minutes, 29 April 1932.

[54]*Ibid,*. 22 November 1935.

[55]*Liverpool Daily Post*, 8 January 1936.

Liverpool "the best-equipped grain port in the country."[56] Although other ports had built granaries in the interim, this monster, which attracted little attention at the national level, reinforced that claim by increasing the total storage capacity by nearly fifty percent. It would be the last new granary needed in the port until the appearance of very large dry bulkers mandated the construction of the silos at Royal Seaforth.

It might be thought that building a huge granary intended to berth quite large ships in Brunswick Dock, with all the attendant problems of the Pluckington Bank as described in chapter six, was foolish. Yet the curious fact is that Unbound Worked-up Paper C116[2], which is where complaints about ships grounding while bound for the silos would be found, is with just one exception silent. This may have been because some of the vessels had discharged part of their cargo at mill berths in Birkenhead, but the whole purpose of building the Brunswick Silos where they were was to enable direct discharge from ocean-going vessels. In the absence of strident and repeated complaints, we have to assume that it was largely successful.

Despite the problematic nature of the Board's original investment at Birkenhead and Waterloo (and of a smaller and earlier one at the Birkenhead Composite Warehouse), its long-term policy of risking money on the initial innovation in order to encourage a new trade, followed by stepping-back and allowing others to continue it, seems to have been both wise and successful. It may be thought particularly so in the case of the Birkenhead mills. The approach for the first dockside mill had been made by Robert Buchanan early in 1892, when he suggested converting the old "Composite Warehouse" adjacent to the Corn Warehouses and in use as an extra grain warehouse, into a mill. The Warehouse Committee suggested it would be willing to rent him some land on the other side of the Dock Road, and the customary negotiations began, with the draft lease being approved by the Board on 17 February 1892. It was, by Board standards, quite a generous document, allowing such perks as toll-free use of the dock railway and toll-free operation of barges within the Birkenhead system. The Board remained benign, as well. In 1898 the Company asked for two improvements, a small jetty to increase its barge-handling capacity and an extra silo. The jetty was built by Lyster's men at a cost of £1390, for which no recharge was made, and the silo was built by a contractor for the Board who then rented it to Buchanan at six percent per annum of what it had cost. The following year it put in a new railway connection for which it required an "acknowledgement" of £1 per annum. When the Company wanted a *pied-a-terre* in Liverpool, an appropriated berth at Victoria was forthcoming. Victoria was only a coastwise dock, but it was conveniently situated for handling barge traffic, particularly cross-river grain from Waterloo.

---

[56]*Corn Trade News*, 10 February 1932.

There were times when the members could be difficult, even by the standards of a body which traditionally drew up extremely mean-minded leases, and one really cannot see the sense in some of the Board's exchanges with the Grain Storage Company. That does not alter the fact that during the period considered here, Liverpool and Birkenhead together became and remained the second-largest milling centre in the world and the pre-eminent corn port of Britain. Conversely, corn became by far the largest trade in the port.[57] The loss of some potential traffic to Hull and Manchester was little more than a slight irritation, and Liverpool's place in both the trading and the milling of corn remained unchallenged. The Board must have been doing some things right, and one of those seems to have been the policy of providing or assisting in the provision of facilities for new trades to encourage others to join in. Certainly its midwifery of Buchanan was conspicuously successful, even when measured against potentially comparable successes in grain storage and the petroleum trade.

Petroleum

The early development of the petroleum trade in Liverpool was extremely rapid and is fairly well documented.[58] It was initially driven largely by the demand for lamp oil, which was considerably cheaper and cleaner burning than any alternative – including gas until the invention of the incandescent mantle. The first purpose-built storage for petroleum products was the "casemates" at Herculaneum: sixty artificial caves for the storage of barrels, and the first bulk facilities were opened in 1891 immediately to the south on land from the former Parkhill estate which the Board had purchased in 1884. There is nothing to prove that it was purchased with a view to being used for a petroleum facility, and indeed the reasons for its purchase are complex and certainly included a wish to forestall the possibility of any kind of competing port venture.[59] While members knew that the land would eventually be useful in the seemingly endless expansion of the port, there were no specific purposes mentioned in the discussions leading to the decision to buy.

---

[57]In 1937, for example, grain imports amounted to 2,109,000 tons while the next largest import, petroleum products, was only 869,000 tons.

[58]See A. Jarvis, "Bulk Petroleum Importation into Liverpool before 1930," in Jarvis (ed.), *Foul Berths and French Spies. Essays on the Port of Liverpool, c. 1800-1930* (Liverpool, 2003), 46-55.

[59]A. Jarvis, "Land Policies in the Port of Liverpool, 1857-1930," *International Journal of Maritime History*, XIV, No. 1 (June 2002), 115-133.

It could be just a matter of luck that it proved to be a highly suitable site for the new trade. It was adjacent to the existing Herculaneum casemates, a location chosen because it stood at the southern extremity of the Dock Estate, and the only housing nearby was occupied by working-class people of no political influence. Should it prove successful, the Board owned plenty more land to the south, enabling easy expansion. Herculaneum Dock had been excavated out of a shallow slope, with the result that to the east and south sides there was a small artificial cliff, allowing construction of the bulk facility at a higher level, always a desirable feature as it allowed all the raising of oil to be done in the process of discharging the ship, and everything thereafter was gravity fed. In fact, it was so well suited, and the Board well enough acquainted with the growth of the trade, that even in the absence of proof we might assume that it could not have been *that* lucky.

The trade on the Liverpool side of the river was never in crude oil; right to the end (in 1967) Parkhill and its extension at Dingle received refined petroleum products. There was no large area of cheap land remote from housing suitable for building a refinery, so when crude importing began in 1923 it went to the new Stanlow Oil Dock on the Manchester Ship Canal, and it stayed within the Port of Manchester until the supertankers outgrew the facilities and moved downstream to the MD&HB's jetty at Tranmere, opened in 1960.

Over the years, the balance between the different products handled on the Liverpool side shifted with changes in technology. As mentioned above, lamp oil initially was the leading product, but in the 1890s the oil lighting trade began to suffer. The incandescent gas mantle appeared, providing a much cleaner and more intense light than even the best oil lamps, and the introduction of the pre-payment gas metre meant that people whose credit was not good could nonetheless now have gas lighting. And an even more formidable rival was on the march. There had been a number of small private companies providing purely local supplies of electricity for lighting since 1883, but in 1896 the City Council purchased the Liverpool Electric Supply Company, and in 1899 began construction of two very large (for their time) power stations. The consumption of electricity for lighting and power multiplied five-fold between 1896 and 1902, and the price per unit (for lighting) fell from 7.5 to fourpence. These were not happy times for the men who sold lamp oil.

Two new markets were just around the corner, with a third not too far away. "Motor Spirit" had in the early days been virtually a waste product, usable as a cheap cleaning fluid; indeed, at some refineries it was simply burned off to get rid of it.[60] Enormous improvements in motor car design occurred in the first

---

[60]It even had a period of use as a cleaning agent in ladies' hairdressing salons, remaining fashionable after the dangers had become (fatally) obvious. See Great Britain, Parliament, *Parliamentary Papers* (*BPP*), 1910 (Cd 5175), XLIV, 609, Report of the

decade of the century, with high-performance luxury cars becoming acceptably reliable (as proven by "reliability trials," which were races in all but name, held over distances of hundreds of miles) while expertise in the manufacture of components improved rapidly. Moreover, the cost of car ownership was declining sharply. As cars got faster and ran on pneumatic tyres, they raised large clouds of dust from traditional water-bound limestone roads, creating a demand for binding agents for "tarmac" - one of which was prepared from the sticky gunge left over in the oil refinery after the desirable middle and upper fractions had been removed.

For some historians it has long been a favourite sport to accuse the Royal Navy of undue conservatism when faced with a new technology, and while this may be sustainable in some cases, it is not in the case of oil firing. In 1905-1906 the aged battleship HMS *Mars* was refitted as an oil-burner, the first large naval vessel to use the new fuel, and the epoch-making HMS *Dreadnought* of 1906 was equipped to burn either coal or oil. In terms of performance and labour-saving there was no contest between the two fuels: the reservations expressed in naval circles related to the higher cost of oil and its world-wide availability in time of war, a problem which had already been solved in the case of coal. The Royal Navy had carried out considerable research into oil-burning equipment, and it is at least arguable that it was the first "shipowner" to come up with an effective and reliable system. Be that as it may, *Mars* and its smaller forebears pointed the way to a large new market for petroleum products.

What we now call "semi-diesel" engines had been around for some time, and Dr. Diesel was granted his patent for a full compression-ignition engine in 1892, but for some purposes they were only slowly adopted. There were but few diesel road vehicles before the late 1920s and very few motor ships before the Great War. Stationary engines were more successful, though the Board's experience with them at Gladstone Graving Dock was not entirely happy. But it was fairly clear to engineers that the thermal efficiency of the diesel engine would entice efforts in continuing development to diminish its disadvantages and that it was in a sense the prime mover of the future. Perhaps its first victory was in becoming virtually universal in British submarines from *D1* of 1910 onwards.

The purpose of this digression into the uses of petroleum products is to emphasise that the members of the Board did not need the powers of Mystic Meg to know a good prospect when they saw one, and that is why they were willing to invest in the Parkhill estate with the port's first bulk-oil facilities, consisting of four storage tanks, a barrel-filling plant and "filling stands" for delivering oil to road or railway tanker vehicles. As in the case of the grain trade, however, what the Board was providing was in effect a "special opening offer;" after that, the oil companies had to provide their own. The work had been paid for from

---

Departmental Committee on Petroleum Spirit, 20.

revenue and appears to have cost about £29,670, and the marginal receipts in dock dues were of the order of £3000 per annum.[61]

As might be expected from the outline of the growth in uses of oil, it was not really until after the war that the big returns on the investment began. In 1914, world tanker tonnage had been under three percent of world total tonnage, but by 1921 this had risen to over seven percent. Oil companies now began to invest at Parkhill. Anglo-Mexican had rented No. 1 Petroleum Depot, and rented additional land from the Board on no less than three occasions within the accounting year 1919-1920, worth a total of £5369 per annum. The same year both Anglo-American and British Petroleum rented significant plots and started building tanks on them, taking the rental income alone, irrespective of dues, to over £10,000. This turned Parkhill, initially a heavy loss-leader, into probably the best investment since the establishment of the Board, with a return comfortably exceeding the bank rate. Unfortunately, there do not seem to be any surviving figures for dues income (a possible explanation for this is suggested below) and estimation is difficult due to the fluctuations in rates, but as a rough indication, dues would be of the order of £200 on the contents of a 3000-ton tank, or about £650 per filling on the four original tanks alone.

This was a time when Pluckington Bank was being troublesome, as related in chapter six, and in particular there were knock-on effects at Herculaneum in the shape of siltation caused by impounding pumping for sluicing and hence fluctuating water levels in the docks through the drawing of sluicing water. Nor was the entrance itself necessarily always operable to its full nominal depth. These were not the circumstances in which to foster the most lucrative new trade the Board had developed in decades, and the outcome was the construction of Dingle Jetty, opened in 1923, where the Pluckington Bank did not reach and the scour of the ebb was likely to prevent it extending. Technical information on the jetty is hard to find. Although there are a considerable number of drawings of Parkhill and Dingle, virtually all of them showing shore installations, no written description of it has yet been found. The reasons are not hard to find: jetties such as that are simple structures, routine jobs possessed of little interest to an engineer and certainly not causing him to feel a paper at the Institution coming on. It was paid for out of revenue, and the jetty itself appears to have cost only £16,514. The whole project, which included a second, smaller jetty for tank barges and coastal tankers, two roads and minor works, totalled about £81,000. Separately accounted for was some relatively complex work to provide rail access to the jetty.

It was clearly a success, but once again we have trouble finding the exact figures, and the most convenient source, the *Petroleum Times*, is little help as it

---

[61]These figures are tentative because they are derived from, rather than stated in, NML, MAL, MD&HB, *Annual Accounts* and Revenue Statements, various years.

aggregates landings of oil for the whole of the port, including some berths on the West Float, Birkenhead, which handled mainly lubricating oils. Furthermore, it gives them in gallons, when everything else is in tons, and conversion is haphazard in the absence of a stated specific gravity.[62] The Weekly Returns (see below) are a problematical source, but for present purposes it is sufficient that they show rapid growth of trade: in its first full year it earned £14,740, or eighty-nine percent of its entire capital cost, and by 1926 it was handling some 20,000 tons per week, representing about £1400 in dock dues alone. Not surprisingly, this led to decisions in 1928 to extend the jetty in two phases of 198 and 700 feet, respectively, allowing it comfortably to berth two of the largest tankers in service, which at this time were about 500 feet long. The barge jetty had been completed in 1927.

The reader will have observed that there are not nearly as many numbers here as in some other chapters, and where they are given they are often as single figures, not time series. The reason for this is the obscure nature of the only systematic internal record I have found. The tabulated Revenue Returns, which are so useful in earlier periods, run out in 1924, and the best substitute which can be found is the Weekly Returns which were submitted to the Docks and Quays Committee. Not only were their contents not transcribed into the minutes of the Committee but even their existence is not mentioned; they are to be found only in the Committee Papers sequence. Extracting them is a tedious process, and when completed gives us only the tonnage of oil shipped, not its origin or destination, which makes it impossible to calculate the dues income. Normally, material "hidden" in this way sets the historian's nose twitching, looking for some failure or corruption needful of concealment.

At Dingle the situation is reversed. The jetty opened at very low cost and immediately produced large and rapidly-growing revenues; dock dues alone reached £1400 per week by 1926. Meanwhile, back in chapter two, Gladstone was grinding on to completion at huge expense. The total cost of the original Dingle Jetty works out at about £33 per linear yard of quay, while a prewar single-story transit shed (the shed alone, not including the quay on which it stood) cost about £230 per yard and was most unlikely to match the throughput of a jetty berth, which discharged vessels at the rate of about 300 tons per hour.[63] To put this into context, 300 tons per *day* was a poor, but not disgraceful, rate for a general cargo quay with only a single-story shed.

The superb new three-story sheds rising at Gladstone could do far better than that, and so they should, for they cost more per linear foot than the whole

---

[62]As a very rough guide, petroleum spirit was typically a little below 0.7, crude about 0.8 and kerosene, fuel oil in between the two.

[63]*Petroleum Times*, 8 July 1922.

cost of Dingle Jetty. The shed on the south quay of Number 1 Branch cost £549,215. The Board was still involved in large borrowings to complete Gladstone, £1,101,129 in 1925-1926 and a further £672,549 in 1926-1927, for example. Comparisons like those above would not make the raising of that money easier. Furthermore, they would exacerbate the complaints from the merchant, coasting and short-sea interests both within and without the Board to the effect that far too much money was spent in the interests of the owners of large vessels and not nearly enough in those of smaller ones. Having worked on the MD&HB archive for thirteen years, this remains the only example I have found of the Board covering up success.

Much of this might have come to nought had the Board not engaged a second time in investment to encourage new trade. It is well-known that the Great War provided a huge stimulus to the development of petrol-driven vehicles and that large numbers of these were sold to the civilian market soon after its end. The largest volume petroleum product handled at Dingle was fuel oil, but the fastest-growing sector of the trade was in petroleum spirits, packed in those square two gallon cans one sees in motoring museums. In 1920, however, the Anglo-American Oil Co. initiated the modern system of filling stations with underground storage tanks filled by road tankers.[64] Petrol, in other words was going to travel to Britain in bulk, and then be discharged, stored and delivered in bulk. It posed very serious fire risks compared with previous methods and other products, but the Board grasped the nettle and constructed the first of the new storage tanks for the trade. The first cargo of bulk "dangerous petroleum" was landed in 1925. As with the casemates and the first tanks at Parkhill, this was sufficient to tempt the initial customers, after which others followed and paid for their own installations.

The timing was perhaps cautious, the first discussions having taken place as long ago as 1920, but it fitted in well with the beginnings of long-distance road haulage mentioned in chapter three. At this time the majority of commercial vehicles were petrol-driven, all but the smallest of them giving a mileage per gallon in single figures. This, too, was the period when the motor bus, in addition to providing minor or rural services, began to represent a credible threat to the former dominance on major urban routes of electric trams.[65] And again they used

---

[64]D.F. Dixon, "Petrol Distribution in the United Kingdom, 1900-1915," *Business History,* VI (1963), 1-19.

[65]The number of trams in service peaked at about 14,000 in 1921, remained at that level until 1931 and dwindled to 9000 by 1938. Some entire systems, like Birkenhead's, were abandoned in the 1930s. The number of motor buses and coaches rose from 40,000 in 1926 to 53,000 in 1930. Perhaps the most interesting indication of the growth in use of goods road vehicles is that their numbers continued to increase by a

large and thirsty petrol engines, which was made easier by the fact that for much of the decade petrol retail prices were distinctly low.[66] These and other developments were reflected in a nationwide increase in imports of refined petroleum products, which remained fairly steady in the early 1920s, rose from 3.9 million tons in 1925 to 7.1 million tons in 1930 and remained almost depression-proof, slipping by only about four percent in 1931 but then resuming its inexorable rise.[67] Perhaps the Board had been over-cautious and perhaps the timing was more luck than judgement, but as it happened, the new facilities came on stream at exactly the right time to benefit from a period of particularly rapid growth. In 1919 Thomas Newell had reported to the Works Committee that there would be an enormous demand for fuel oil in the immediate future. He may not have identified exactly the right product, but the Board's investment policy in encouraging the trade was to prove an enormous success. Coinciding as the growth in the trade did with the economy measures described in chapter five, it made a significant contribution to the port's emergence from the Depression in surprisingly good shape. The percentage of the Board's dues income spent on interest peaked at about seventy-five percent in 1934, a year which the Chairman could describe in his *Annual Report* as "uneventful but marked by slow improvement," and then fell sharply to only minimally over fifty percent in 1938. This was in part the result of falling interest rates, but both dock rates and town dues (which were charged on the cargo only and are therefore a truer indicator of the state of trade than is aggregate income) both showed growth, of six and 5.6 percent, respectively. It had again been Liverpool's good fortune, as in some of the great financial crises of the nineteenth century, that when one of its key trades (and it was often cotton) was doing badly, another trade was growing strongly to soften the impact.

---

couple of percent per year throughout the depression. B.R. Mitchell, *British Historical Statistics* (Cambridge, 1988), 557.

[66]Dixon, "Petrol Distribution."

[67]Mitchell, *British Historical Statistics*, 271.

Chapter 8
Into the Modern World?

It will be recalled that in 1890 the Board appointed a Special Committee to look into the general conduct of its business. In some respects this can be seen as one of a series of enquiries which set out to simplify over-complex and ineffective administrative procedures which were correctly perceived as unnecessary additions to operating costs.[1] Its enquiries went further than that, however, and considerable attention was paid to the quality of service the Board offered its customers.

Prior to this enquiry, the Board had been becoming increasingly conservative in its policymaking and increasingly self-satisfied. With the occasional glitch, tonnages and revenue kept rising, and the rise was as marked in the higher-value trades like premium cargo-liner services as in the lower-value ones. The decline which occurred in Liverpool's share of the whole port services market was mostly due to a relative decline in low-value trades, notably coal exports, and hence of less significance to the port's real position than might at first appear.[2] Even the Board's enormous capital debt could become a source of perverse pride, for there were not many organisations to which individual investors would lend £18,951,678, which was the total authorised borrowing of the Board in 1890, marking another reason why the Board could consider itself, if not an unique institution then at least a very special one. The previous year G.F. Lyster had read a paper on the recently-completed dock extension programme at the Institution of Civil Engineers, and it is clear that even that gathering of big spenders was impressed with the scale of Lyster's work and the funding available to him.[3] The programme was indeed a huge one, whose size alone might justify some measure of pride on the part of the members.

---

[1]For further details see A. Jarvis, "Attempts at Reducing the Expenditure of the Port of Liverpool, 1836-1914," in P.C. van Royen, *et al.* (eds.), *Frutta di Mare: Evolution and Revolution in the Maritime World in the 19th and 20th Centuries* (Amsterdam, 1998).

[2]As Jackson pointed out, the meteoric growth of coal export from the South Wales ports and elsewhere, distorted the "market share" figures for the whole industry. See, G. Jackson, *The History and Archaeology of Ports* (Tadworth, 1983), 116-117 and 141-142.

[3]G.F. Lyster, "Recent Dock Extensions at Liverpool," *Minutes of the Proceedings of the Institution of Civil Engineers*, C (1889-1890), 2-114.

It was, however, already becoming clear to some members that those works, only finally completed in 1888, had been based on a serious underestimate of the depth of water needed to keep the port competitive in those trades using the largest ships, particularly those on the North Atlantic, which had been Liverpool's most important trading area for many generations. But the Board was a body in which the Chairman received enormous respect and correspondingly wielded great power, and between 1876 and 1889 the man who wielded it was Thomas Hornby. It was Hornby who did his best to pretend that all was well at the New North Works, defending Lyster and the Works Committee against the attacks of Harold Littledale and others who thought the design and execution of the works both costly and ineffective.[4] In 1889 Hornby died, and the enthusiasm with which the local press greeted the election of Alfred Holt as his successor rather suggests that Hornby had the reputation of holding back development in the port. His support for Lyster had extended to the very unpopular refusal to dredge the bar, mentioned in chapter six, and, as recounted there, less than a year was to elapse before Lyster grudgingly began dredging. Within another year he was in semi-retirement. Did he jump or was he pushed? The record is unclear.[5]

The purpose of this sketch of events outside the timeframe of this book is to suggest that in 1905 there were two deep-rooted strands in Board thinking: one was that of the survival of the pre-1890 "we know it all" party, the other that of the men who took the ideas of 1890 thoroughly on board and were a little worried by what they saw around them. Characteristic of the former group was an initial refusal to have anything to do with the formation of a port authorities association at a time when the "customer problem" referred to in chapter five was at its height.[6] A less consequential, but perhaps more telling, decision when considering the desirability of appointing a General Manager was that there was no point in enquiring into the experiences of other port authorities which already had one because there was nothing to be learned from the practices of other

---

[4]The quality of the finished works was, as usual, excellent, but it was the cost which caused complaints, often on the assumption that it would be cheaper to use contractors than to keep work in-house, which may have been true but owing to the complexities of the work was never positively proven.

[5]One thing which is clear, however, is that this was *not*, as formerly believed, a move to further the career of Lyster Jr. Lyster Sr. had placed his town house in Belvedere Road in trust for his son, but his will left him not a penny, so presumably there had been a great falling-out.

[6]National Museums Liverpool (NML), Maritime Archives and Library (MAL), Mersey Docks and Harbour Board (MD&HB), Worked-up Papers (WUP) 101, contains two sequences of letters, one from Swansea in 1880-1881 and another from Hull between 1887-1889.

ports.[7] Liverpool knew best. It is, of course, obvious with hindsight that it did not, and we find from time to time refreshing glimpses of a more progressive attitude beginning to creep in and even in some respects take hold. This process was not helped, however, by the establishment of the Port of London Authority in 1908, which was closely and deliberately modelled on the Mersey Docks and Harbour Board. Imitation is the sincerest form of praise, and it was taken as such.

Probably nowhere was this insularity and need for change more obvious than in the attitude towards competition. From pretending that there was none, the Board had to learn that in some traffics it was losing market share to other ports, notably Hull (grain and oilseeds) and Southampton (passengers). This was only a relative decline, and not yet serious, but it did demand attention. The question then was what to do about it, and the immediate reaction was to turn again to the solution which had apparently served the Board well in the past: provide superior facilities. It was not just a question of following up the 1898 programme with the Gladstone project. The strange arrangement at Canada Entrance, a persistent source of trouble and expense, was modified yet again, eliminating the unnecessary Brocklebank Entrance. The passages at Canada and Brocklebank had been built in several stages over a long period of time, each to the size then considered (rightly or wrongly) appropriate for the largest ships of the day, and the Brocklebank improvement programme, completed in 1908, standardised them at the depth of Canada Lock and a width of ninety feet and added nine acres of water space.[8] The inadequate Carriers' Docks were converted into a very large graving dock and a deep branch dock. The usable quayage and the maximum size of vessel able to use it both increased considerably, and some efficient two-story ferro-concrete transit sheds (added after the completion of the main scheme) made a further improvement to the ensemble.[9] Proceeding just in advance of these extensive (and technically difficult) works was the construction of Canada No. 3 branch, completed in 1906, which added more than seven acres to what were now the best deep-draught berths in the port.

But the Board had followed that path before and, as in previous cases, the new facilities were enthusiastically taken up and used, but by existing, not

---

[7]The discussion of the Special Committee's report is in *ibid.*, Discussions at Board 15, 22 and 29 January 1891.

[8]This was fourteen feet six inches below Old Dock Sill, representing about thirty-one feet of depth on the sill at mean high water.

[9]For details of the scheme, see W.H. Jones, "The Brocklebank Dock Improvements on the Mersey Dock Estate," *Transactions of the Liverpool Engineering Society*, XXXIII (1912), 164-200; and J.F. Ramsbotham, "Graving Dock Construction, Dealing with the New Brocklebank Graving Dock," *ibid.*, XXVIII (1907), 45-116.

new, customers. All that happened was that the firms which previously enjoyed the best facilities moved into the new best facilities, allowing everyone down the line the chance of improvement in a long and sometimes complex series of moves. Ultimately, all that was happening was a net transfer of tonnage and revenue from the worst berths to the new ones.[10] The figures that prove this come, of course, from the Board's own archives, but it seems that their significance was not fully recognised. They show that simply offering a highly superior service at a rather high price would not necessarily cause new customers to flock in. The story would be repeated at Gladstone and with the Clarence Modernisation Scheme described in chapters two and three.

Clearly there was a marketing problem, and for purposes of marketing a service at that time we can identify several components in the process. The most elementary is knowing what the customer wants or needs, and it might be thought from the foregoing that the Board was ineffective in market research. Perhaps it was, but it is difficult to understand how or why. Its entire *raison d'être* was to provide the services the customers required, and the vast majority of its members were elected specifically as representatives of one group of customers or another. As we have seen in chapter five, if the Board did not know customers' needs or act upon that knowledge, there was no shortage of individuals or trade associations ready and able to point this out. More to the point, as mentioned in chapter 4, the average number of memberships of trade associations held by members of the Board rose from an average of slightly under 0.5 towards the end of the nineteenth century to nearly 1.5 in the early twentieth. This might be seen as an inbuilt market research mechanism and certainly should have functioned as one.

Furthermore, when a major project was envisaged, in addition to receiving marketing information available through those channels, the Board deliberately canvassed opinions in what we now refer to as a consultative process. These varied in both extent and effectiveness, partly because they served an important secondary purpose of trying to forestall Parliamentary opposition to the bill which would be the end product of the process.[11] By and large it would be fair to say that the construction and major modernisation projects described elsewhere in this book were based on a reasonably thorough investigation of what the customers wanted - or at least of what they *said* they wanted. The emergence of people like Dr. E.L. Corthell as almost specialist shipping futurologists is a good indicator that while customers might know what they wanted at the time of

---

[10]This issue is considered in the context of the coastal trade in chapter three and explored in detail in A. Jarvis, "Think Big? Returns on Different Scales of Investment in the Port of Liverpool, 1873-1905," in Jarvis (ed.), *Foul Berths and French Spies. Essays on the Port of Liverpool, c. 1800-1930* (Liverpool, 2003), 20-34.

[11]See chapter two for a description of the pacification of opposition to Gladstone.

asking, they had little idea of what they would want even in ten years' time, which in some cases could be less than the time taken to see a major project through from its first outline design to completion.[12] Ideally, the Board needed to build for a working life of about twenty years as a "top dock" before obsolescence set in, which could mean designing thirty years ahead. This brought its own special hazards, in that almost by definition, the further ahead one tries to guess the future, the less likely it is that the guess will be correct, and there is some evidence that less ambitious schemes for the improvement of existing facilities were more likely to be sound investments than gigantic new-build or re-build projects.[13] As suggested in chapter two, it is possible that in the case of Gladstone the Board actually built further ahead of shipping developments than was really necessary.[14] If that is so, then clearly a cheaper job would have released income for dues cuts additional to those made in the late 1920s and early 1930s, which might have won the port new trade from its competitors. The Board, however, was generally of the opinion that since Liverpool could never be a really cheap port it should concentrate on quality of accommodation and service, a view strongly endorsed by Sir Alfred Jones (who was never a member of the Board) when giving evidence before the Royal Commission on the Port of London. Adam Kirkaldy did little to help deflate the corporate ego: "It was Liverpool that first taught the world how a great commercial river and port should be run."[15] Even that falls far short of this breathtaking demonstration of complacency from Board member Hugh Rathbone: "The Port is well-known to Shipowners and Shippers as the biggest and best port in the world. It is perfectly well understood that Steamers must come to Liverpool."[16]

Despite the apparent ignoring of the figures and the happily atypical view of Mr. Rathbone and a small handful of others, those members of the Board who

---

[12]Clearly Gladstone is not a fair example to cite because of the problems caused by the war, but the works under the Acts of 1891 and 1898 each took ten years. The "futurology" issue is discussed in more detail in A. Jarvis, "The Port of Liverpool and the Shipowners, c. 1910-38," *The Northern Mariner/Le Marin du nord*, XII (2002), 23-40.

[13]This is explored in more detail in Jarvis, "Comparison."

[14]Though with hindsight the result of *not* building Gladstone as far ahead of actual need as it was would probably have been a need to build a very large new break-bulk dock during the 1950s, the consequences of which would have been dire indeed.

[15]A.W. Kirkaldy, *British Shipping: Its History, Organisation and Importance* (London, 1914), 519.

[16]NML, MAL, MD&HB, Discussions at Board, 20 January 1921. The context was his opposition to passing the advertising estimate for the following year.

one might loosely term "progressive" had identified a course of action they felt should be adopted whose vulgarity would have horrified their forebears in the 1880s. They were going to forsake the assumption that steamers "must" come and actively seek after business. They did not have a good starting point because while the Board might be respected for its financial security or for the huge amount of business it brought to the city and the region, in a ballot for the most unpopular institution in Liverpool it would probably have beaten the workhouse into second place. The Board's chosen methods of public relations at the time were perfectly rational but open to misinterpretation – and sometimes to jealousy. One of their major tools was what we would now generically dub "junketing," often involving *Galatea*, a vessel which was nominally a tender but in reality a steam yacht of some luxury. This, of course, did not delight the gentlemen of the press, whether local, national or specialist, partly because they did not necessarily get invited. The Board's view was simple: it entertained people who might bring it business, which journalists would not.[17] The potential customers in Liverpool were all perfectly well known on the shipping and mercantile networks of the port, and those people knew what services the port had to offer; if they were in doubt, they knew who to ask and were unlikely to be influenced by a newspaper. In that sense Rathbone was right: it was his assumption that the old ways would remain effective that was wrong.

All of these themes had come together in an incident in 1893 which forcefully illustrates the background of ill-will against which this chapter stands. The Board's annual "Cruise of Inspection" had originated as just that: a trip to allow Members to verify that outstations like lightships, lighthouses and telegraph stations were indeed in the good order which had been reported by their officers. Its purpose evolved first into a mainly social occasion and then, through the invitation of increasing numbers of local decision-makers, into a networking session. Provided the weather was reasonably calm, the members and their guests could enjoy a thoroughly pleasant occasion enhanced with good food and fine wines: it was a time for *bonhomie*. A few select members of the press were normally invited. What happened on the 1893 cruise is inexplicable. John Brancker, the Chairman of the Board, chose this most inappropriate of occasions to launch an intemperate attack on the alleged ignorance and malice of the press. The following day, 1 July, *Commercial World, The Liverpool Post* and the *Liverpool Review*, none of them especially noted for previous hostility to the Board, quite understandably responded in kind.[18]

---

[17]The Board also gave journalists plenty of other reasons to hate it, including a perceived excessive secrecy with regard to matters of little or no obvious sensitivity.

[18]NML, MAL, MD&HB, Newscuttings I, B64.

Right down to the time of its demise, the Board rarely saw any need to try to appear "nice" to the public at large. It was not necessarily an ungenerous body; there were regular donations to local charities, tenders of the Board's fleet were lent out free of charge for transporting VIPs to and from the training ships on Prize Days, and various little derelict corners of the Dock Estate accommodated pleasure boats free of charge "on sufferance." The Board was in the main a generous, if autocratic, employer, though it is typical of its attitude that "compulsion" in the shape of the Workers' Compensation Act and the National Insurance Act made them less rather than more generous towards disabled ex-employees or their widows. In more recent years the amount of information provided for school children taking their first halting steps in economic geography was considerable. These things, however, were deliberately not publicised for two reasons: the decision-makers whom the Board might wish to influence knew about them already, and it feared that if any generous little gesture appeared in the local press "the others would all want one too."

The terrible Public Relations gaffe mentioned above was far from unique. On another occasion it sued Robert Cook, a fisherman, for trespass, in that he did not stay on the marked path when crossing an empty plot of land the Board owned at Tranmere to get to the foreshore, and it obtained an injunction restraining him from doing so. When he was found doing it again, it pursued the matter and on 23 September 1898 Cook was gaoled for contempt of court. After ten weeks, the Board agreed to his release if he swore an affidavit to the effect that he would not do it again, he would not allow his sons to do it and he would endeavour to prevent anyone else over whom he had any influence from doing it either.[19] Large notices were erected reminding walkers that they were there on sufferance, and fishermen were issued with permits which they were required to carry and show when requested. This, be it remembered, was on a vacant plot which had been bought as a preemptive purchase to prevent a repetition of problems it had suffered at the hands of land speculators.[20] Public access to the waterfront has long been an issue on which the Board has made itself unpopular with the public at large; it might be said that was a price to be paid for the prosperity the docks created, but that did not apply to other lands in its ownership.

The view that public opinion did not matter had been deeply damaging in the nineteenth century because there was one thing the Board wanted very much from ordinary citizens and, more particularly, the local authorities which represented them, and that one thing was that they should not object to the

---

[19]*Ibid.*, Unbound Worked-up Papers (UWUP) T107, vol 1.

[20]A. Jarvis, "Land Policies in the Port of Liverpool, 1857-1930," *International Journal of Maritime History*, XIV, No. 1 (June 2002), 115-133, gives more detail of the background to such incidents.

Board's bills in Parliament. Select Committee enquiries were a large and often avoidable expense, involving the employment of, among others, two or three barristers and a string of expert witnesses. In the Board's legal files we can find successive drafts of the same proofs of evidence, indicative of the very substantial amount of staff time which went in as well. As the Board's success with the 1906 Bill showed, it was possible to come to agreements without that expense, and the adoption of a more opene and conciliatory attitude yielded dividends. Seven further bills were promoted between then and 1938, and all but one passed unopposed.

Figure 10
Trans-shipment of Textile Machinery from L&L Boats

Source: National Museums Liverpool.

One of many problems in the running of a port in the years just before the Great War was that a number of large and expensive facilities had been built for what one might broadly term "non-economic" reasons. It is fairly generally agreed now that the construction of the Manchester Ship Canal was more a matter of politics and local pride than of need. Very large facilities were built at Fishguard, Milford and Immingham for which there was no real need except as a competitive stick with which shipowners could beat the older port authorities for lower dues. There was, in short, considerable surplus capacity in the industry. When, for example, Manchester began to do well at timber importing it was largely at the expense of Fleetwood, which might well have been put out of business had it not been for the rapid expansion of the steam trawling industry there. In these circumstances, restricting marketing activity to the old gentlemanly methods of the commercial grapevine and tailoring the service to market needs would no longer suffice; it became necessary to adapt the techniques of the consumer market in the shape of advertising and the employment of people who were in effect salesmen. The recognition that port services were a buyer's market, and that there was no reason at all why steamers *had* to go to Liverpool or anywhere else in particular, came rather late to the Board.

In the early stages of researching this section I was surprised to find the Board engaging the services of an advertising agent in the 1880s, which might be thought to disprove much of what is written above.[21] In fact, the role of the agent was to publish formal notices of such things as new issues of dock bonds, the drafting of a new Parliamentary bill or an obstruction to shipping caused by the Board's engineering works. The word "advertisement" was being used in its archaic meaning, which related solely to providing information, not to persuasion, which is just as well since the latter type of advertisement had gained an evil reputation at that time.[22] Indeed, the Board long continued to be castigated for being an accessory to bad taste by renting out poster space on the floating stage.[23]

The first definite move towards more aggressive marketing of the Board's services was the appointment of an "outdoor canvasser," first suggested at the Board meeting of 3 November 1910. In the discussion which followed it emerges that most other ports already had at least one, and the matter was followed up in further discussion on 29 December. The title "canvasser" was dropped, and it was decided to appoint a "Traffic Agent" in the General Manager's office, with assistants in London, Birmingham and Bradford.[24] Mr.

---

[21]NMGS, MAL, MD&HB, UWUP A119.

[22]This was largely through misrepresentation, which was widespread especially in advertising patent medicines, but bulk telegram advertising – the equivalent of today's "junk fax" – was perhaps the most offensive practice. For a general history of advertising, see T.R. Nevett, *Advertising in Britain. A History* (London, 1982).

[23]See NML, MAL, MD&HB, UWUP A120.

[24]There was by this time a considerable literature on salesmanship, advertising and allied topics, but not much evidence that Board members were familiar with it. Perhaps they dismissed it because much of it was published in America. In the event, the Yorkshire office was initially located in Leeds, only moving to Bradford in 1922.

Holt, in many ways a progressive, spoke strongly, even bitterly, about "touting" and the wastefulness of competition. On the latter point he was right, but unfortunately the competition, particularly from Manchester, which was very active in "touting" at the time, would not go away, and on 4 January 1911 the Board confirmed the appointment of Robert Walker as Traffic Agent. The wording of his "job description" in the staff records is informative: he was "to obtain information with regard to the movement of traffic in which the Board may be interested." The appointments to branch offices followed, though not at a pace suggestive of great urgency, despite the fact that most of the other principal ports already had agents in Birmingham and London. The appointment of the Birmingham agent was confirmed by the Board on 25 January 1912, and the stipulation that the whole cost including the Agent's salary, the wages of an office boy and the rent of the office should be "not exceeding £300 per annum" suggests that it was unlikely to recruit a high-flyer.[25]

The London agent was not appointed until 25 June 1914 and then only on an initially experimental basis for one year, which seems remarkable in that the Board already had rented premises at 13 Victoria Street, and the upkeep of a London office had first been taken on in 1830. It was not, however, a traffic agency but a *pied a terre* for the solicitor when in London on Parliamentary business, and a gentleman named Wainwright was paid an honorarium of £50 per annum to "mind the shop" when no one from Liverpool was there. Such was the volume of Parliamentary business that his role was probably not very demanding. For a time the costs were shared with the Chamber of Commerce, but in 1874 the solicitor "informed the [Parliamentary] Committee that he had taken rooms for the use of the Board at No. 13 Great George Street, Westminster at a rental of £100 per year."[26] There is a curious reference to a Mr. Arthur H. Aspinall being engaged to "assist in the conduct of the Agency" and an equally irritating passing mention that the "Subscription" be discontinued after 30 June 1882. Taken together these suggest that the Board was involved in two quite different offices, and the *Annual Accounts* for 1881 confirm it, with separate entries for the rent of "Board's Office, London" and "Grant to Liverpool Office, London."[27] The only likely reason for the latter expenditure was the general furthering of the commercial interests of the port, and indeed any other purpose would probably have been *ultra vires* under the terms of the Board's Consolidation Act of 1858.

---

[25]NML, MAL, MD&HB, Discussions at Board, 4 January 1911; and 25 January 1912

[26]*Ibid.*, Parliamentary Committee, minutes, 12 November 1874.

[27]The relevant minutes may be found in NML, MAL, MD&HB, UWUP L91.

The Board now rented a succession of premises around the Victoria area until 1914, when it rented the lower ground floor of Billiter House, Billiter Street. This suggests a fundamental change in role, for all the previous premises were within easy walking distance of the Houses of Parliament, and one of them was in fact owned by Rees and Frere, the Board's Parliamentary agents. Billiter Street, on the other hand, runs between Leadenhall Street and Fenchurch Street, and was presumably intended to be conveniently sited for seeking to raise capital in the City and for a little touting of the most gentlemanly kind.

The London office virtually shut down for the early part of the war but reopened on Tuesdays to Fridays from 6 December 1917, and the speculation above is strengthened by the move to ground floor shop premises in Leadenhall Street at a rent of £400 per year, the highest yet paid. The shop window contained only some elegant plain mahogany panelling, since window displays were still very lower class, but there was display space inside for displaying models of Liverpool docks. The turn of the century might not have brought about a change of heart in favour of more modern marketing methods, but it seems that the war did, for it was not just the style of the London office which changed.[28]

The first problem at the end of the war was not, as had often happened at the end of previous wars, a recession. On the contrary, the problems of congestion during the brief boom were considerable and have been mentioned above. Once the boom collapsed, matters took on a very different complexion. The Board had occasionally taken part in trade fairs: a "Panoramic Painting" of the port which was exhibited at the Shipping, Engineering and Machinery Exhibition at Olympia in 1923 was an updated replacement of one purchased for the St. Louis Exhibition of 1904. The Board would continue to refuse invitations to exhibit,[29] but now it now threw itself with some enthusiasm into planning a major exhibit at the British Empire Exhibition of 1924, including commissioning a huge working model described as a"model showing the Dock Estate at Liverpool and Birkenhead with moving liners, ferry boats and oil tankers in the river and moving trains on the Overhead Railway."[30]

The General Manager was later obliged to excuse the model makers from making the trains work when it was pointed out to him that the scale was such that they would be 0.07 inches high. Even so, the whole model came out at forty by eighteen feet, necessitating a large "stand" at a cost of £684 in rental. It

---

[28]The first definite mention of a window display involved the "Panoramic Painting" (see below) on 10 December 1931. *Ibid.*, UWUP Q17[1].

[29]Including some, like the International Tobacco Trade Exhibition of 1925, which appear to have offered good marketing opportunities.

[30]NML, MAL, MD&HB, Trade Committee, minutes, 1 February 1923.

next fell foul of the organisers in that its base did not comply with regulations and an approach was made to Professor Reilly to see if he would re-design it. He would not, but he provided them with a man who did, this time satisfactorily. The Board's troubles were not over yet: it realised that for the stand to do any good it needed to have a member of staff there all the time, a commissionaire to "guard the model" and a telephone. These were among the reasons why an initial estimated cost of under £2000 rose to £5325 at the end. By 24 April 1924 the stand was completed but was still lacking its electrical supply, and shortly afterwards the General Manager reported to the meeting of Chairmen that there was a bar immediately opposite their pitch, whose customers were using the Boad's stand for seating accommodation. They agreed that this was unsatisfactory and authorised him to try to do something about it, in which he failed. When invited to exhibit again the following year the Board declined on grounds of the poor treatment it had received, and the next major effort was for the International Colonial and Maritime Exhibition at Antwerp in 1930 [31]. For this it commissioned a working one inch = thirty foot model of Gladstone Dock, with real water, lock gates which opened and shut, a tide in the river which rose and fell and "14 or 15 vessels," of which three were to "move about." The estimated cost of the model was put at £1550-£1750 and the total cost of exhibiting at £4000-£5000.[32]

The Traffic Agents exhibited a good deal of "drive." At their meeting of 2 March 1923 they suggested that the Board ought to publish a "shipper's guide" to the services of the port to be used as a handout at the Wembley Exhibition and at subsequent meetings suggested "a general information booklet consisting principally of pictures" and another one broadly similar but "on rather more technical lines" and then went on to a "Monthly Sailings List." The Board allocated the "ball park figure" of £4000 for literature for distribution, and the General Manager eventually ordered 300,000 "Souvenir Folders, 25,000 Souvenir Booklets, 25,000 copies of the *Guide to the Port*, 105,000 postcards of black-and- white aerial views and 180,000 of three-colour views. The total estimated cost of this was £2416, and I have found only one passing mention of any earlier consideration of providing "literature for distribution."[33]

The Traffic Agents had until now been "invisible" in the published *Annual Accounts*, though we know that from at least 1920 there had been an

---

[31]This account is assembled from the contents of *ibid.*, UWUP Q17.

[32]Passed by the Board on 12 September 1929.

[33]In the same discussion as Mr. Rathbone's attack, Mr. Willmer stated that "Last year we published a book which I believe has done the Board a great deal of good in foreign countries" (NML, MAL, MD&HB, Discussions at Board, 20 January 1921), but further details, both of the book and of its distribution, have yet to come to light.

"advertising budget" presumably hidden away in one of the catchalls like "Sundry Charges at Dock Office," and we also know that some members at least were already well aware of the need to advertise: "We will all agree that business today is very largely the result of advertising the facilities of the various ports and cities ...indeed it is the spirit of the age." Mr. Willmer probably did not really think that Rathbone and Holt would agree, but the payment for the new panoramic picture was agreed.[34]

A much bolder initiative, which is also "hidden" in the *Annual Accounts*, was the Board's decision at its meeting of 29 June 1922 to send "a Board's representative to Australia and New Zealand for propaganda purposes," and four weeks later it decided that the representative should be Colonel Hawkins, Principal Assistant to the General Manager. He left in September and was away nine months. The Board was obviously well pleased with his extensive reports, as there was no opposition to the proposal to pay him an honorarium of 500 guineas when he returned. His final report is a substantial document, laden with names and addresses of potential trading contacts.[35] It is also surprisingly dispassionate, providing facts and figures on Australian trade and industry without making any strong suggestions as to what trade might be gained there, though he emphasised the importance of wool. He encountered traffic agents for the ports of London, Manchester, Hull, Southampton and Immingham but got the impression they were not much use and did not recommend that Liverpool should appoint one, suggesting instead that the Board should send printed information to the agents of Liverpool shipping companies there. Perhaps his key point is that what the Australian trade needed was more demand for services from the UK end. Nevertheless, at the national level, trade with Australia was one of the few whose lowest point in the Depression did not dip below 1913 values, so his somewhat muted enthusiasm was not misplaced.

The purpose of the trip was clearly to try to penetrate London's dominance in trade with the Antipodes, and four years later Hawkins was on his travels again, this time to the International Cotton Conference in Egypt to try to steal some of Manchester's trade.[36] In 1928, Watson Rome, the London Traffic Agent, visited Australia and New Zealand, and in addition to generally drumming up business he was to urge interested "local houses" to lobby for direct cargo-

---

[34] *Ibid.*, 4 November 1920.

[35] Colonel Thomas Henry Hawkins, CMG, entered the Board's service as "an additional assistant to the General Manager" in 1919 and was promoted Principal Assistant in 1921. *Ibid.*, Staff and Wages (SW) 1/23. His report is in Management 1.

[36] Manchester had not done much damage to Liverpool's preeminence in American raw cotton but had done well in the Egyptian trade.

liner services to Liverpool, an instruction which makes it quite plain once again that the intention was to attack London's trade.[37] Once again, lobbying at the UK end was also undertaken.

In the *Annual Accounts* for the year ended 1 July 1924 we find the first entry under "General Charges" for "Traffic Agents Expenses, Advertising etc." in the sum of £9259 11s 6d, and in the following year of £8563. Both these figures presumably included costs incurred at Wembley, for that entry ran at around £5000-£6000 for the next few years, and its decline also masked a marked increase in advertising in the specialist press. In 1923 the Board placed regular advertisements with *Lloyd's List,* the *Chamber of Commerce Monthly Journal* and *The Journal of Commerce*. In 1924 it added *Fairplay, Syren and Shipping* and *Shipping World* to the list, fairly predictable choices if the policy was generally to advertise more. What is more interesting is that it also added three Antipodean publications, the *Pastoral Review, Meat and Wool* and the *New Zealand Dairyman*, followed the next year by *The Grazier's Review*.[38] It is clear that Colonel Hawkins' visit, perhaps reinforced by networking at, or in the process of preparing for, Wembley had convinced the Trade Committee that there was a worthwhile market in carrying the agricultural produce of Australia and New Zealand. It does not appear that the Board advertised in the trade press in other dominions of the Empire.

It was almost certainly in 1923 (the antecedents are unclear) that the practice began (mentioned in passing above) of holding regular monthly meetings of the Board's Traffic Agents, apparently the first such meetings of groups of "officials" of which formal minutes were kept.[39] They exhibit another unusual characteristic in that the Agents, being in direct and frequent contact with the customers, represented their views to the Board (via the Trade Committee) and at times showed signs of "going native." At a meeting on 12 January 1923, for example, they raised the issue of "the lack of uniformity in the classification of goods in the tables of Dock Rates and Town Dues, the schedule of master porters' charges etc." The dues structure of the Board had indeed remained complicated, to the point where it was difficult for a potential customer to work out what the overall cost of handling his cargo would be. It is highly questionable whether there was still any need for dock rates and town dues to be calculated separately anyway, and certainly no reason why the tables of rates should be different.

This was a bit too radical for the Board, which agreed in principle but postponed the consideration of the matter because the whole business of charges

---

[37]NML, MAL, MD&HB, Discussions at Board, 13 September 1928.

[38]*Ibid.*, UWUP L136^4.

[39]*Ibid.*, WUP 144.

was subject to constant change owing to the instability of the market for port services and the intense competition which was prevalent.[40] If this was just an excuse, it was nevertheless an excuse with some substance, reflected in the fact that between 1921 and 1925 tonnage using the port rose steadily from 15.8 to 19.1 million tons while, owing to repeated adjustments to rates and dues, revenue remained almost static.

It is only when one realises that no "give-away" information such as that prepared for the exhibition was previously available to potential customers that it becomes clear just how dilatory the Board had been in matters of advertising even when using the word in its old-fashioned sense. Swimming downstream in a continuous and seemingly irreversible growth of world trade, it had been possible to leave the customers to go and buy the *Shipping World Yearbook* and find out for themselves; in the postwar world it was not. So far as the record goes, it appears that 1923 thus also marked the first appearance of any printed information which the agents in the branch offices could hand to enquirers.

Rome was not built in a day, so when the *Canadian Export Pioneer* offered a deal under which it would publish a "four-page special" about the port in return for the Board paying £32 for a two-page advertisement, the Trade Committee would have none of it, despite the *Pioneer* cannily pointing out that it had already published such a supplement on the Port of Manchester.[41]

There is another peculiar strand to the apparently fairly swift change in the Board's advertising policy so soon after the war. It might be attributable in part to personalities, because Thomas Rome succeeded Sir Helenus Robertson as Chairman in 1919 and Lionel Warner succeeded Alfred Chandler as General Manager and Secretary in 1920. Warner came from the railway industry, which had made extensive use of advertising long before the ports industry, so we may assume that he exercised at least a little informal persuasion on the members. Whether he could in effect have initiated a completely new policy is another question. Some new members were elected to the Board, but not in atypical numbers and, as explained in chapter five above, the gerontocracy factor in the Board was not falling but rising. Exactly what generated the members' support for statements like those of Mr. Willmer, quoted above, is not entirely clear, but that it happened is not in doubt.

From the members' *Pocket Yearbooks* we learn that from 1913 a new standing committee existed, to whose minutes reference is made above, namely the Trade Committee, whose principal duties were to retain existing custom and

---

[40]For example, the Traffic Agents' Meeting of 5 April 1923 heard rate cuts made by Trafford Park Cold Storage Ltd. described by United Cold Storage as "ridiculous and incapable of being maintained."

[41]NML, MAL, MD&HB, Trade Committee, minutes, 29 January 1925.

attract new users. It is difficult to find anything important or effective that it did in the first ten years of its existence, after which it suddenly sprang into action. What is even more curious is that I have failed to find any record in the Board minutes or elsewhere of the establishment of the Trade Committee. Although it was a full-blown standing committee, in theory though not in practice as powerful and important as Works or Finance, its minutes were never bound up in the manner of those of every other standing committees, even including the Parliamentary Committee, which only met "as required" rather than every week, but were kept instead as a "Worked Up Paper" (No. 144). This is peculiar from the other side of the question as well, in that WUPs are compilations of material from various sources, sorted according to subject, and No. 144 is the only one of over 200 of them that is not. Furthermore, in 1933 the Trade Committee disappeared from the *Yearbook*, and we find the Traffic Agents reporting to the Traffic Committee.

This definitely does not indicate that the Board had given up on advertising because it was not bringing home the business. In 1930, some years after the other main ports, the first publicity film was produced and was certainly approved of as useful by the agents; indeed, the London agent was so impressed with it that he asked for a "Duoscope Automatic Daylight Cinematograph Cabinet" in order to be able to show it in the London office.[42] Various small "gift" publications, like calendars and blotters, were produced and distributed, and the print run of the *Port of Liverpool Handbook* kept getting longer, reaching 4000 in 1930. The year ended July 1930 was marked by the considerable expenditure on the agents' expenses and "Advertising the Port" of £10,521, and despite the successive waves of economy measures described in previous chapters, by 1934 the same head of expenditure had fallen only to £5977 – a modest enough cut when compared with laying-up *Leviathan*.

In 1927 the branch agents had been supplied with a Morris Oxford each at an estimated total cost of £500. This in fact means the agents in Bradford and Birmingham only, because as it happens, the Bradford agent had a tyre blow out, causing him to lurch across the road and collide heavily with a lamp post. He escaped with a few cuts and bruises, but the car was a write-off, and its replacement cost was £234, indicating that £500 bought two cars, not three, and this is made explicit when they in turn were replaced with two new ones in 1933. Whether the level of usage changed is unclear, but these two only lasted four years before being described as worn out and replaced with two Austin

[42]*Ibid.*, 26 June 1930. There were two models available, costing respectively £110 and £175, but the minute simply says "Declined." The film was in fact a joint venture with the Corporation and the Liverpool Organisation.

Goodwoods.[43] The point here is that in the dark days of 1933 the Board was willing to buy the agents new cars, recognising the continuing importance of their role.

One form of publicity at which the Board had been quite good was entertaining VIPs or bodies of decision-makers. It had collected some good publicity for its hospitality to the Institution of Mechanical Engineers on the occasion of their Liverpool Meeting in 1891 and to the Institute of Public Health Congress in 1903. The Bradford agent (employed fairly specifically to try to pinch trade in textiles from the Humber ports) was responsible for a visit by the Bradford Textile Association in 1924, and the Board's designs on Australian trade were done no harm by the visit of the Australian Minister for Agriculture and a party of Australian journalists in 1926. In 1927 there was not only the huge party for the completion of Gladstone Dock but also separate visits by the High Commissioners of South Africa in January and Australia in February. *Galatea* was kept busy for a while until it followed *Leviathan* into mothballs.[44]

The question is, of course, whether the Board's new-found adoption of relatively modern marketing methods, and more particularly advertising, actually did any good. It is obviously meaningless to try and put numbers to any appraisal because an estimate of how well it would have done had it not advertised would involve a counterfactual exercise of a particularly speculative kind. The port's share of the total market for port services continued to fall during the interwar years, and some of its traditional trades, notably the export of cotton goods, were in bad shape. The previous ability to identify promising new trades and be in them from the start seemed to desert Liverpool entrepreneurs, though the wool trade exhibited strong growth and petroleum products did well. Grain importing and flour milling expanded strongly during the 1920s. But all that this sequence of statements serves to illustrate (and many more could be provided) is that apart from a small upward trend, with some vicissitudes, in total traffic in the interwar years and a rather larger downward trend in market share, no clear pattern emerges. The good fortunes of the wool trade, however, might reflect the

---

[43]The story of the Board's road vehicles is rather patchy, but may be traced to some degree in NML, MAL, MD&HB, WUP 201.

[44]*Proceedings of the Institution of the Institution of Mechanical Engineers* (1891); *Handbook for the Conference of the Institute of Public Health* (Liverpool, 1903); and NML, MAL, MD&HB, WUP 144/1.

advertising effort in Australia, to say nothing of the visits made there and the work of John MacDougal, the agent in Leeds (later Bradford).[45]

From 1929 onwards, the Chairman's annual report was printed and published along with the *Annual Accounts*, and this document invariably comments on the general state of trade in the past year, but only once in the first ten reports are any figures given which indicate the success or otherwise of the initiatives mentioned above. In the report for 1937 it is remarked upon that in 1925 trade with North America amounted to twenty-five percent of the revenue of the port whereas in 1937 it was only fourteen percent. Australian trade had doubled from four to eight percent, but no remark was made on this, nor any moral drawn; the figures were quoted purely as an illustration of the way that trade changed, with the message that the port must be ready and able to change with it. We are faced once again with a slightly enigmatic silence, yet it is surely no coincidence that on 20 January 1938 Harold Davies was appointed as the Board's agent in Australia at a salary of £1000 per annum, from which he had to fund his own office accommodation and expenses.[46]

It might be thought that the Board's response to the new facts of economic life in the twentieth century was too little and too late. It did, however, change its ways considerably, and the things it did were by no means bad. The key points are the willingness to change, the perseverance to resist the temptation to "cut advertising first" and a perhaps surprising degree of aggression towards the trades of London, Manchester and Hull. This is the real significance of the events related in this chapter. Before the war the Board believed that if only it continued to expand and improve its infrastructure, the ships would come. A situation where there were more berths available than ships seeking them was beyond its experience and imagination. There was, in effect, no new trade to be won, and the only way even of standing still, much less of advancing, was to seek to steal the trade of other ports. Of course, Richard Holt was right in denouncing such competition as wasteful, but this situation had been building up for at least fifty years, and it was a good thing for the port that the majority of members recognised their powerlessness in changed times: the Board now had to compete using not only the Chief Engineer's army of navvies but also complimentary blotters and calendars.

---

[45]John MacDougal was appointed in 1920 at a salary of £400 plus out-of-pocket expenses. It may be significant that by 1940 he was the best-paid of the branch agents on £800 plus expenses.NML, MAL, MD&HB, SW/1/21.

[46]*Ibid.*

Conclusion

On the Eve of War

In the thirty or so years' history of the port considered in this book it is possible, despite the disruption of the Great War both generally and more specifically in the shipping industry, to find marked continuities in the way the port worked. The most obvious of these was the structure of the port authority in that the members appeared to continue to be elected in the same way they had since 1857, and even the changes in the reality of the matter related in chapter four could not be said to represent any striking alteration in the way things worked. The system of Standing Committees remained essentially the same, *pace* the arrival and disappearance of the Trade Committee, and so did the system for appointing their senior staff. The power of the General Manager grew noticeably, but while the growth might have accelerated in the 1920s it was also in a sense a continuity in its own right, which can be traced right back to Daniel Mason's unsuccessful attempts to gain a little power here and there from Jesse Hartley back in the 1840s, passing through the sometimes childish squabbles of Miles Kirk Burton with the Lysters, father and son. What did change was the General Manager's relationship with the members, particularly after Warner was appointed in 1920. Once his feet were well under the desk they began to allow him responsibilities which previously they reserved to themselves, an expanded role which is well illustrated in the dealings with the Unemployment Grants Committee mentioned in chapter three. This had the important effect of freeing up time for consideration of the strategic issues which were the proper business of the Board and its committees; the discussion of trivia shows a marked reduction.

Another marked continuity is that of the Board's debt, which continued at such a high level that expenditure on capital improvements might have its timing determined not by the Board's choice but by the ruling rate of interest – or by its best guess at what interest rates would be in the future. This was particularly difficult in the early stages at Gladstone, as the bank rate typically changed about five or six times a year, and in 1910 it changed nine times. There was, of course, no question of ever paying the debt off, and another continuity is that when the Board had a windfall such as the money from the sale of the Clarence Dock site to Liverpool Corporation it never occurred to it to apply it to the reduction of its debt; it had to be spent on port improvements, which were still seen as the major incentive to shipowners to use the port. What did change, however, was that the headlong pursuit of provision for the largest ships afloat finally ceased with the completion of Gladstone, and the Clarence Modernisation was the largest and most ambitious programme of works specifically for the coastal trade in the entire history of the port.

A high priority for any surplus money that might be available was to make cuts in the rates of dues. This was not new in principle: reductions in dues had been employed to damage competition from Birkenhead in the 1840s and Manchester in the 1890s, but the practice in the interwar years was to tinker with rates on a much more flexible and selective basis, according to the twists and turns of trade. Some of Liverpool's leading trades, notably cotton imports, remained both fairly weak and unstable, and nationwide exports of cotton goods fell to barely over one-fifth of 1913 levels. The cuts in dues (and the occasional minor increase) had a certain "smoothing" effect on fluctuations and may well have been a significant factor in the port's relatively successful performance in the worst of the Depression. This was aided, as in the past, by the diversity of Liverpool's trade, with some improvement in the Antipodean wool trade and a marked growth in the grain trade.[1] The Dingle oil jetties were almost certainly the Board's most "profitable" investment ever, despite attracting only a small percentage of the trade nationwide. Conversely, Liverpool had comparatively little exposure to the marked decline of the coal export trade, which caused very large reductions in the trade of those ports which depended more heavily on it. As Gordon Jackson has pointed out, this decline, when expressed in aggregated national tonnage figures, has masked a marked growth in some other trades, just as the coal trade distorted our ideas of the competitive position between ports during its rapid growth in the late nineteenth century.[2]

Considering the generally weak state of shipping and trade, the Board made a surprisingly large investment in improvements to its infrastructure in small matters as well as large, and this was in part enabled by another element of continuity, namely the practice of periodically appointing a special committee to investigate and report on the general conduct of business. The first of these to do any noticeable good was held in 1890, but the three in our period were definitely constructive efforts that cut out a good deal of waste, which is perhaps surprising when we note how conservative the Board could be in other ways. It was, in some respects at least, more progressive than those who sought to urge progress upon it, and certainly more so than some of its customers. Money could not be

[1]Francis Hyde's analysis of Liverpool's declining competitive position *vis-a-vis* other ports is for present purposes not particularly helpful because it tends to deflect attention from actual growth in Liverpool to the greater growth elsewhere which caused a fall in market share. The fact is that Liverpool handled 3.6 million more tons of shipping in 1938 than in 1913, an increase of almost twenty percent.

[2]Gordon Jackson, *The History and Archaeology of Ports* (Tadworth, 1983), 141-142.

spent everywhere simultaneously, but minor projects such as electrifying the antiquated pumping apparatus at Princes and Canning graving docks cost only a few thousand pounds and produced a saving in wages which would last as long as the need for the docks. The bridge renewal programme might have eased traffic, but it also eliminated the nightmare maintenance costs of the ageing swing bridges which were replaced and reduced the number of men needed for operation.

There was one conspicuous exception to this picture of a Board with its back to the wall doing surprisingly well, and that exception was the continued adherence to the Jesse Hartley philosophy of vertical integration. That this was a wise policy when first adopted is beyond question, as is the fact that by the mid-1960s the Dock Yard was as inefficient as one could possibly imagine and then some.[3] The question is what went wrong and when? I have previously suggested that the first signs of decline in the Dock Yard appeared in the 1880s, but these were very preliminary. Within our period, the withdrawal of "the bosses" to the new Dock Offices in 1907 clearly diminished the self-esteem of those left behind, and the persistent attacks by Board members who thought that it was cheaper to put work out to contract continued the process. But the success of the Department at Gladstone graving dock seems to have wiped out some unhappy memories in the boardroom - until the managerial disaster at Alfred Entrance. But as emerges in the main text, the eventual Board view of that was that it was largely its own fault for running a sloppy system and, bolstered by Unemployment Grants Commission funding, it continued to undertake successful major works in-house. If there was to be any hope for the future it lay in carrying out a thorough modernisation of the Dock Yard, suggested in 1913 (and 1919, 1925, 1939 and 1956), which might both have rationalised its use of space and put some pride back into the people who worked there, not least by giving them decent equipment with which to work. Getting rid of the rubbish might have helped as well. Alternatively, it could have built an entirely new and much smaller Dock Yard to undertake only that work which could not be appropriately put out to contract. To give an idea of the scale of the operation, the Engineer's labour force peaked near the beginning of our period at a figure in excess of 10,000, and even as late as 1967 was still over 3600.

Another aspect of integration was the operation of the Board's fleet of vessels, and this grew considerably during our period, most conspicuously by the acquisition of the 200-ton floating crane *Mammoth*. Nobody seems even to have considered the question of "privatising" them, and the only way in which they featured in economy drives was that some of them were temporarily laid up. The

---

[3]In 1967, the consultants Urwick Orr and Partners Ltd. investigated the running of the Engineer's Department. It found labour utilisation never reached fifty percent and was often below twenty percent.

only conspicuous disposal of any in-house functions was the relinquishing of the leases of the Board's quarries at Kirkmabreck in 1908 and Dinorben in 1911, and by 1938 its "fleet" looked like the basis of a particularly seedy maritime museum.[4] It was a continuity it could have done better without.

During the 1930s the modernisation programme began to slow down as well. While all the hydraulic pumping stations except for Waterloo appear to have been electrified by 1938, all the old impounding stations were still driven by reciprocating steam engines of a fairly old-fashioned design and low efficiency, using steam at only 100 pounds per square inch, as were some of the graving dock pumps.

If we partially repeat the imaginary trip on the Overhead Railway recounted in the Introduction, what are the visible changes? Glancing backwards as the train emerges from the Dingle Tunnel, just before reaching Herculaneum Station, the extensions to the petroleum depots and the more extensive railway lines are obvious. Because of the lay of the land, the Dingle jetties cannot be seen, but vessels at them can. In most other respects Herculaneum is unchanged and carrying little traffic: the majority of the vessels are there for bunkering or to use the graving docks. From Toxteth through to Brunswick the only conspicuous change is the addition of the very large silo, and on again from there to Canning little has changed apart from the completion of the works which were in hand in 1905. The remains of Georges Dock Passage have been modernised, with a new wide-span transit shed, and the upper part of Mann Island is somewhat tidied up by the building of a new motor garage. Manchester Dock and Chester Basin have been infilled, but the buildings around their sites are still in place. The former empty site behind the Dock Offices is now occupied by the Mersey Road Tunnel ventilation tower.

At Princes Dock the east sheds are new, and the graving dock has been converted into a wet berth with quite a large shed. The Central Dock Yard is still there and still as untidy as ever. Over at West Waterloo are the first signs of the construction of the new river entrance, and immediately beyond Waterloo we encounter the first of the new Scherzer bascule bridges and the fruits of the Clarence Modernisation programme at the remodelled Victoria, Trafalgar and Clarence Docks, with smart modern facilities for coasters, though the dominant impression is that of the large new power station which paid for the improvements. A well-timed glance up Saltney Street will allow a glimpse of the extension to the Colonial Wool Warehouse. From here to Sandon Dock there is little change, but Sandon, Canada and Huskisson still accommodate large passenger vessels and now have large two- and three-story concrete transit sheds

---

[4]The oldest was probably the derrick flat *Canada*, built in 1828, but the water and stores boat *Mersey No. 37* was sixty-one years old, grab dredgers nine and thirteen were built, respectively, in 1884 and 1893 and the crane barge *Atlas* in 1886.

equipped with electric cranes on some of their quays.[5] Langton and Brocklebank have not changed a great deal apart from better railway provision, but the appearance of Alexandra has been transformed with the construction of another large granary and an equally impressive cold store. Hornby remains much as before, but the greatest change of all lies immediately beyond: Gladstone Dock, which was barely at "fag packet sketch" stage in 1905, is complete and doing good business. In 1938 it is still as good a facility as may be found in any impounded port in the world.

The thirty-three years which have elapsed have been marked by major improvements, most carried out in adverse circumstances, of which Gladstone and the Clarence Modernisation are obviously the most important, and some of the relatively minor improvements, such as the new shed at Princes, are perhaps more effective than they look. But the obsolete stretch between Canning and Kings remains unaltered, and throughout most of the south docks little or nothing has happened since the completion of the 1898 programme. Astonishingly, the lower part of Mann Island is still a virtual wasteland.

Gordon Jackson, in a chapter heading in his *History and Archaeology of Ports*, characterised the period 1919-1980 as "Stagnation and Decay," which for most ports viewed over the whole of that period is clearly fair. What we must try to judge is how far the stagnation and decay advanced between 1905 and 1938 in Liverpool.

The physical appearance of the estate was probably a fairly true indication of its history over those thirty-three years, though the device of the imaginary journey enforces the omission of significant works on the Birkenhead side. Those were years with a continuing need for the port to adapt to changes in its trade, most particularly in the continuing growth of the average size of vessel. There were qualitative changes as well. The strong performance of the grain trade is reflected in new silos, while the decline in coal exports has left Bramley-Moore and Wellington unmodernised. The "failure" to modernise the Canning-Kings group reflects the fact that, for the first time in its history, the port had spent more than ten years with more berths available than the customers needed. The Board, and even the government, had been aware of this and had undertaken some deliberately counter-cyclical improvements, but the "bottom line" remained, as it always had been, the ability of the Board to borrow enough money at a price it could afford, and particularly with the decline in dues income since 1929 this

---

[5]The number of hydraulic cranes (Liverpool side only) fell from 214 in 1905 to 188 in 1938, while the number of electric ones rose from zero to 117, all but twenty of these being from Huskisson northwards. National Museums Liverpool (NML), Maritime Archives and Library (MAL), Mersey Docks and Harbour Board (MD&HB), *Yearbooks*, 1905 and 1938.

ability fell well short of what was needed to bring all of the Estate up to anywhere near the standards of the best parts.

But that is a very negative view. The physical state of the port was probably better overall in 1938 than in 1905; certainly the proportionate area of land and water which was totally obsolete was smaller. The *status quo* of 1905 rested on a history of growth in trade which, with the occasional hiatus such as that caused by the American Civil War, had continued for many generations. The achievement of the period 1905-1938 is that in absolute terms as distinct from market share the port arguably did better than between, say, 1875 and 1905 and did it in circumstances of acute difficulty.

If the physical state of the docks is a useful measure, it still cannot be the only one. It was essential for the Board and its principal officers to be able to recognise the need for change in other matters as well. Here, the record is mixed. On the one hand, it took on new ideas, such as the need to advertise; on the other, it never, right down to the time of its final collapse, came to terms with the need to be liked: it remained capable of behaving in an apparently arrogant manner. This was particularly foolish because in many important ways it had cast off the old "we know best" type of arrogance, and when the corporate belt needed tightening it laid up *Galatea* with fairly good grace. Almost any problem which came along it faced up to and, unlike its forebears, was willing to make hard decisions.

There are some aspects of the Board's business which seem to belong to another age. A trifling, but perhaps telling, example is that when a new member was elected, the invariable expression in the minutes was "in the room of Mr Hodgson, deceased," a form of words dating back to the days of the old pre-1857 Dock Committee, which, along with several other archaisms, was still in use in 1938. Reading the Chairman's annual reports for the mid-1930s one still gains an impression of an intensely old-fashioned organisation. It may be, of course, that this is conditioned into us by our knowledge of what would happen after the war, when booming trade once again, as before the Great War, papered over large cracks until about 1966, whereupon the whole edifice collapsed.

But that is unfair: we should allow the simpler criteria for success to stand. During our thirty-three years the Board faced serious problems over which it had little or no control, and it left the port generally in a better state than it found it.

Appendix

Port Efficiency: A Short Methodological Discourse[1]

Calculations or estimates of port efficiency appear in many places in this book, and they would not do so if the author did not believe that they are of some value in the understanding of port development. Such figures are not a recent construct, as are some which have been calculated by historians since the days of "The New Economic History" of the 1960s. On the contrary, they were often calculated in the nineteenth century for presentation to Parliamentary Select Committees when applying for powers to build, extend or modernise a particular dock or group of docks. That, however, does not necessarily make them trustworthy, or release us from the obligation to use them in an intelligent manner. The fact that "new" figures have been calculated from a normally reliable source such as the Board's Revenue Statements, does not mean that they need not be interrogated just as critically as other sources.

If, for example, one measures efficiency (as the Board often did) in terms of tonnage handled per linear yard of quay, then to make valid comparisons it is essential to know what was being handled and how it was handled at the time in question. Discharge of timber during the period covered in this book differs dramatically from timber discharge in the latter days of sail because the sailing ships moored bow-on to the quay, hence the length of their holds became the notional midship cross-section on the basis of which cargo density was calculated. Their efficiency in terms of tons/yard was higher than that of more modern vessels, but, of course, their efficiency in tons per water acre was less. The use of tonnage per square yard of quay was a less popular yardstick at the time, but that too would show a comparatively low efficiency for a dock in which timber was important because of the large "standing ground" required, to say nothing of the space needed to handle butts up to forty or so (and exceptionally up to seventy) feet long.

Conversely, imported bulk grain was not an especially "good" traffic in terms of revenue per yard, but, standing in silos rarely less than eighty feet high, it looked absolutely superb if assessed in tons per square yard of quay. The coastwise trades could, as mentioned in chapter three, achieve surprisingly high efficiencies in tonnage per yard, and they could do it in extremely obsolete docks,

---

[1]Much of the material in this appendix is outside the time frame of the main text of the book. This is simply because the small body of published work employing the techniques discussed relates largely to the 1870-1914 period.

but because of dues concessions, their performance in revenue per yard was generally poor. Here, as in other cases, the extreme example is Dukes Dock, considered too humble even to appear at all in the Revenue Statements. As mentioned above, one customer alone, Spiller, carried 300,000 tons of grain in and out of Dukes in 1930, giving it the astronomical, and totally meaningless, efficiency figure of 132,158 tons of goods per water acre. The average for the whole Dock Estate was about 25,000 tons.

There is a serious difficulty with graphs of efficiency of individual docks over time, and it does not vary significantly according to which of the various methods is used to calculate efficiency. Broadly speaking, efficiency graphs are the same shape as total trade graphs. There seems only one obvious explanation why rising trade was almost synonymous with rising efficiency and that is that there "must have been" excess capacity waiting to be taken up. Yet we know perfectly well from narrative sources that this was only rarely the case, and that for most of the period between 1873 and the end of the postwar boom the port was suffering various degrees of congestion (mainly in the sheds and on the quays rather than in the water), sometimes as the result of berths being unavailable through engineering work, at others simply through obsolete facilities or the unusual volume of trade (as in chapter five). It must also be remembered that poor trading conditions, as well as favourable ones, could cause congestion in the form of vessels waiting for cargo, laid-up or even, in extreme cases, abandoned. In 1907, for example, the Harrison Line complained to the Docks and Quays Committee on 6 March that its vessels were having trouble reaching their appropriated berths because of Elder Dempster vessels laid-up five abreast at *their* appropriated berths nearby, and that was a year in which trade was only poor rather than dreadful. It seems unlikely that there was sufficient spare capacity in virtually every dock to allow extra trade just to be soaked up.

It might be suggested that the whole method is false and that the peaks in the graphs simply represent the point beyond which no more goods can be shifted and that because we have not taken ships' time in dock into account and the dimensions of the dock are not variable, then the maximum volume of trade must necessarily equal the maximum efficiency, and the graphs are thus a truism. But this is not so, for the tonnage per water acre would be free to rise by double or treble-berthing whereupon the congestion caused on the quayside would result in at least a levelling off, possibly a diminution, of the tonnage per yard of quay. In other words, peak volume is not the same for each of our four ways of measuring, yet the curves all still follow broadly the same patterns. Yet another problem is that in the ways set out in chapter five, the *effective* dimensions of docks are, in fact, variable as well.

If that sounds like a message that there is much work yet to be done here, it gets worse. The impression is readily gained from Bird or from Jackson that the construction of docks which set out to provide full-length side-on berths for steamers resulted in a larger quay length-to-water area ratio, through their being either long and narrow or built with a spine and branches, often in the shape of a letter E. They were so built, but it is not true that their quay-length to water area ratio was necessarily increased.

Alexandra and its branches, opened in 1880 to a classic E-shape design, impounds fifty-eight square yards of water per yard of quay, while Albert, a type-specimen of the almost square sailing-ship dock, impounds only fifty-six. The real space-waster turns out to be the maximum length of ship needing to be swung to or from the berths. In 1910 the berths most recently modernised for handling transatlantic passenger liners were at Sandon, where the ratio was seventy-three square yards of water per yard of quay, the highest in the port at the time, and at Canada and Huskisson. At berths intended for more modest-sized steamers, as at Toxteth and Harrington, the ratio is much more favourable at fifty and forty-eight, respectively. In 1901 the ratios in the main system varied between twenty-six at Carriers Dock and seventy-three at Sandon; roughly speaking they varied directly with the ship size envisaged when the dock was designed. (or, in the case of Sandon, last re-designed) This is emphatically confirmed by the customarily extreme case of Dukes Dock, built for flats in 1773 and barely altered since, which had a ratio of just nine.

It seems at first sight, then, that there ought to be a relationship between the efficiency of a dock calculated as above and its water area-to-quay length ratio. The larger ratios were encountered in docks designed for larger ships and, as is well known, larger ships were more efficient; that is why people bought them. This relationship was plotted in the confident expectation of finding, if not a neat rising curve, then at least a vague rising trend. The resulting graph shows that there is in fact no relationship between the two sets of figures, and even more curiously, it does not matter which efficiency figure we use: the curves vary, but all four are equally chaotic. There has to be a reason, and the only one which seems even mildly plausible at present hinges once again on the different trades and roles of specific docks, or, to put it another way, to the exceptional complexity in real life of what we are now trying to reduce to neat simple numbers.

To begin with, in 1910, 50,442 vessels entered or left the port, excluding "Tenders, Tugs, Flats, Fishing Craft and Smaller Craft than Schooners."[2] Some of these would be fairly homogeneous in the nature of their use of the docks – a Harrison liner was for those purposes much like an Elder Dempster liner. Those two vessels would have approximately similar cargo densities, would be mostly discharging primary produce and loading manufactured general merchandise using similar equipment at similar berths. Their sizes would not be wildly different. Others, however, varied in all these respects and others as well. Very

---

[2]National Museums Liverpool (NML), Maritime Archives and Library (MAL), Mersey Docks and Harbour Board (MD&HB), *Report of the Acting Conservator*, 1911.

little work has been done on vessels in ballast which must also affect the statistics. Some docks were used extensively for moorings for unladen vessels waiting to enter graving docks. Then, as in several places in the main text, we ought to try to take into account the berthing and unberthing of the Board's own vessels - several thousand movements per year.

Lewis Fischer has attacked the problem of measuring efficiency from a completely different direction, by working from the dock registers to calculate turnaround times.[3] The registers are a particularly bulky and awkward source, and he is the first to extract anything meaningful from them. He is able to show measurable improvement in performance, of eleven per cent between 1870 and 1890 and of three percent between 1890 and 1910, for the ocean-going trades. In the coastwise steamer trade he finds a rise of seven and six percent, respectively, over the same periods.

He then seeks confirmation of these trends from the dock registers, and the trouble with the figures he obtains is that such is the intractability of the registers that it took an enormous amount of work to construct the sample on which they are based from the records for just one dock, namely Huskisson. He concludes that turnaround times for coastwise steamers fell by twelve percent between 1870 and 1890 and by eight percent between 1890 and 1910. As it happens, the history of Huskisson at this time was fairly typical of the Estate as a whole in that it underwent some radical changes during the period under consideration. In 1870 it was still a fairly modern dock as built in 1857, while by 1890 it was definitely ageing. Under the 1891 Dock Act, Canada and Huskisson were deepened and the large new entrance at Sandon constructed to serve them. The result as revealed in the Revenue Statements is pretty much what one would expect. When Huskisson opened in 1857, its largest trading area in revenue terms was the United States, and the revenue grew steadily from £26,751 in 1857 to a peak of £80,074 in 1881.[4] It then dropped precipitately to £69,853 in 1882 and £27,278 in 1883. Meanwhile, the trade with the east coast of South America rose from £13 in 1881 to £16,909 in 1883, and Mediterranean trade also rose, albeit more slowly, from £24,804 in 1881 to £50,559 in 1884. The reason for these fairly sudden changes was the opening of Alexandra, which immediately took most of Huskisson's US trade, which dwindled to a trough of £248 in 1897. In 1899 it began to rise again and by 1903 had reached £150,711 because the deepening works were at last complete, providing better facilities than at Alexandra, so the US trade came back again. If we turn to the figures for

---

[3]L.R. Fischer, "Maritime Infrastructure," in J. Armstrong and A. Kunz (eds.), *Coastal Shipping in the European Economy 1750-1980* (Mainz, 2002).

[4]Apart, of course, from the dip caused by the American Civil War.

Alexandra we find, unsurprisingly, that its US trade had fallen, but trade with the east coast of South America and with the Mediterranean had risen.

Why does this game of musical docks have any bearing on the matter? It is because the changes in relative importance of different trading areas introduce a qualitative change in the cargoes and hence an extra variable into comparisons of efficiency (as indicated by turnaround time) over time. To take a very simple example, the rise in Huskisson's US trade from £92,656 in 1902 to £150,711 in 1903 indicates the arrival from Alexandra of the appropriated berths for Cunard and White Star "greyhounds" with their large register tonnages and low cargo densities. Both companies ran large and reputedly efficient in-house master porterage operations to speed up turnaround. Measured by turnaround, therefore, they will look good, while measured by revenue per yard of quay they will look bad, by goods tonnage per yard even worse, by gross register tonnage per yard rather better. To know whether the rise in efficiency is genuine and meaningful we need to know whether any significant changes in typical cargoes lie behind the changes in trading areas which might affect typical cargoes and hence rates of discharge or loading. In the case of passenger liners, cargo discharge was likely to be fast owing to the amount of bulk grain carried eastbound, which, as recounted in chapter seven, was discharged overside by floating elevators, allowing simultaneous discharge of other cargo and hold baggage to the quayside.

There is another factor which needs consideration in the context of Fischer's finding that there was a significant reduction in the proportion of coasters having to wait in the river before being able to enter the docks, and that is the increase of the size differential between ocean-going and coastwise ships.[5] Let us assume for the sake of argument that there was one hour either side of high water when the White Star liner *Cedric* could get in and out. A vessel drawing ten feet less would be able to use the entrance two hours either side. The new Sandon Entrances had about thirty-seven feet six inches of water on the sill at mean high water, which means that between about one and one-half and two and one-half hours before and after each high tide smaller coasters - those drawing perhaps sixteen feet - could come and go unimpeded except by tugs and barges. Thus, an efficiency saving could arise for the coasters entirely accidentally and quite independently of the actual operations of the port. In addition, however, the foregoing remarks refer to Sandon when used as a half-tide entrance. It was in fact a dual-mode entrance, capable also of being used as a fully-locked entrance for vessels up to 165 feet long by forty feet wide, which gave a great deal more

[5]From two to less than one-half of one percent.

latitude at tide time, but only for the smaller vessels.[6] The same increase in size differential came to exclude virtually all ocean-going vessels from several dock entrances, notably in relation to Huskisson, the Salisbury Half-tide entrances to the south and (until they were removed in 1904) the twenty- and thirty-two-foot locks at Brocklebank to the north, where such vessels as could pass them had their exclusive use.

This does not mean that Fischer's choice of Huskisson as an example was either unwise or unlucky. As emerges in chapter five, the only docks which were neither extensively altered nor affected by shifting trading patterns or major engineering works elsewhere in the Dock Estate during the period he considers were those which were totally obsolete and inhabited by the coastwise trade, and mostly the tramps at that - and they, as remarked upon before, often exhibited very high efficiency figures when calculated by number or tonnage of vessels or goods, with Dukes remaining, as ever, the *reductio ad absurdum*.[7] This is because, in a sense, the Board, followed by historians, was using the numbers only as an indirect indicator of what it really wanted to know. The figure for Dukes was not going to convince it that the most cost-effective way to improve the port was to build more docks just like Dukes. *That* was the question to which it needed an answer because it assumed that the future would continue to require constant change as the past had done. The thing it most needed to know was not what could shift the most goods but what would provide a return on investment in order to fund the next cycle of improvements.

Is it, then, worth bothering with these numbers at all when they clearly only present a very partial picture and the amount of labour which would be involved in producing a full picture is inconceivable? Definitely so, for while every attempt so far, not least my own, has weaknesses, each shows us a little more of what was happening, each is a minor refinement of our knowledge, and their inadequacies serve to point to future research agendas. It was, for example, the realisation that Brunswick Half-tide had an efficiency of zero because it was *entirely* occupied by the Board's own vessels and floating plant that led to the approximate calculations of berth availability in chapter five - an issue which had not been previously addressed and which has a very definite bearing on port efficiency. The central problem which faced me when I first started to research the history of the port was to try to find out how the port *really* worked, not how the insiders of the Board had wanted the outsiders, historians included, to think

[6]The eighty- and 100-foot passages were also fully-locked and capable of passing two or three vessels abreast up to a total of 130 feet length.

[7]As related in chapter two, by 1900 the larger and more important coastwise companies were getting relatively good appropriated berths at (for example) Nelson and Princes.

it worked. Owing to the bewildering array of variables which need consideration, the calculation and interpretation of efficiency figures is still a rather blunt tool, but it is a far from useless one, provided always that we use it in conjunction with the others in the box.

Note on Sources

Almost all the unpublished material cited in this book is in the Maritime Archives and Library (MAL) of the National Museums Liverpool (NML), and that location should be understood unless another is given. Within the MAL, most of the material used is from the Mersey Docks and Harbour Board (MD&HB) collection, which should again be assumed unless another collection is named.[1]

Some categories of document in the MD&HB collection are effectively published works, and three of these have been used extensively throughout the book, notably the *Annual Accounts,* the *Report of the Engineer* and (to a lesser extent) the *Report of the Acting Conservator.* Despite their legal status, I have treated them as archive material and cited them in the form given in the list of abbreviations below. The forms of these documents vary over time, but in only one significant respect for present purposes. From 1836, when the first *Report of the Surveyor* was produced, until 1887, the report also contained fairly detailed accounts of the Engineer's expenditure as well. From 1888 and throughout the period of this book it is necessary to use the version kept in the Engineer's office, which contains the printed report bound together with manuscript accounts, MD&HB ENG 1/1-77.

The Minute Books are the prime record of the decisions of the Board and its standing committees. It is important to remember that newspaper reporters were present for the "public business," so the proceedings were normally pretty anodyne. They have extensive and accurate index volumes, but it is often easier and just as effective where only a low level of detail is required to work from the Worked Up Papers (WUPs, see below) and then move to the Committee Files, where further detail is needed.

The Committee Files almost enable us to see issues through the eyes of the members because they contain the documents they had before them at the meetings. Their use, however, is extremely labour-intensive, and in a substantial minority of cases a similar grasp of the members' views on contentious or confidential matters can be gained from the *verbatim* sequence Discussions at Board, which has the further merit of helping us form an impression of what individuals were like. For the earlier part of the period they are extremely useful, but from about 1930 they become progressively "thinner."

[1]A *Guide to the Records of Merseyside Maritime Museum* has appeared in two volumes, the first compiled by Gordon Read and Michael Stammers (St. John's, 1995) and the second by Dawn Littler (St. John's, 1999). While falling far short of a full listing, which would be completely unmanageable, these provide an invaluable introduction to the collections.

The Worked-up Papers are subject-sorted collections of transcribed minutes, other documents and sometimes press cuttings, bound in volumes. In some places I have used these extensively for such purposes as constructing the basic narrative of the building of Gladstone Dock, and where footnotes for such narratives appear thin it should be assumed that it derives from the last WUP cited. Unbound Worked-up Papers (UWUPs) are similar compilations but of smaller bulk or perceived importance.

The Legal classification comprises a large section of the collection: almost all the examples cited in this book come from categories A (the Board's Parliamentary Bills), B (Local Authority Bills affecting or potentially affecting the Board's interests), C (Railway Bills and other matters), E (litigation) and F (real estate). They provide a very fine level of detail, and evidence given for or against bills can often cast some light on gaps in other parts of the collection.

The vast bulk of statistical documents are rather disappointing. The Revenue Statements, which might best be described as a manually-calculated spreadsheet, enable income to be disaggregated on a dock-by-dock basis and become a most versatile tool when entered into a spreadsheet such as Excel. Unfortunately, they end in 1924, and the Steamship Owners' Returns, which start in 1870, end in 1926. From then until 1938 aggregated figures are easily entered into Excel from the Annual Accounts, but disaggregation is possible only by very laborious extraction from the Committee Files.

Newscuttings form a large subject-sorted section of the collection. The cuttings volumes are useful for working out a basic narrative and often contain *verbatim* snippets from meetings. Because the Board was such an unloved body the press could be a bit abrasive, acting as an antidote to the "official view," but they only rarely managed to get their teeth into anything really sensitive.

The Registers of Officers allow the construction of career outlines for salaried staff as, to a lesser degree, does the Register of Established Servants. There is also a good deal of material on accidents.

The Engineer's Commonplace Books contain reports, notes of experiments and memoranda of successes to be repeated or mistakes to avoid. They are often informative on how things were really done, but they unfortunately peter out in the 1920s.

The Management Files are extremely patchy but contain the occasional gem in the latter part of the period.

Bibliography

Andrews, C. "The Somerset Dock at Malta." *Minutes of the Proceedings of the Institution of Civil Engineers*, XXXIII (1871-1872), 352-375.

Anon. *Port Facilities of Great Britain: The Report of the Port Facilities Committee appointed by the Chamber of Shipping, by the Association of British Chambers of Commerce and by the Federation of British Industries*. London, 1924.

Anon. *The Coastwise Trade of the United Kingdom, Past and Present, and Its Possibilities*. London, 1925.

Anon. *The Liverpool Corn Trade Association, 1863-1953*. Liverpool, 1953.

Armstrong , J. "Climax and Climacteric: The British Coastal Trade 1870-1930." In D.J. Starkey and A.G. Jamieson (eds.). *Exploiting the Sea: Aspects of Britain's Maritime Economy since 1870*. Exeter, 1998.

Belchem, J.C. (ed.), *Popular Politics, Riot and Labour: Essays in Liverpool History 1790-1940*. Liverpool, 1992.

Biles, J.H. "The Draught and Dimensions of the Most Economical Ship." *Transactions of the Institution of Naval Architects*, LXXIII (1931), 243-263.

Boyce, G.H. *Information, Mediation and Institutional Development: The Rise of Large Scale Enterprise in British Shipping, 1870-1919*. Manchester, 1995.

Brassey, Lord. "Merchant Cruisers and Steamship Subsidies." *Transactions of the Liverpool Engineering Society*, XXVIII (1907), 30-38.

Brodie, J.A. "Presidential Address." *Minutes of the Proceedings of the Institution of Civil Engineers*, CCXI (1920-1921), 6-29.

Broomhall, G.J.S. and Hubback, J.H. *Corn Trade Memories, Recent and Remote*. Liverpool, 1930.

Cashin, J.A. "Engineering Works for the Improvement of the Estuary of the Mersey." *Journal of the Institution of Civil Engineers,* XXXII, No 7 (1948-1949), 296-367.

Chaloner, W.H. "Charles Roe of Macclesfield (1715-1781): An Eighteenth Century Industrialist." *Transactions of the Antiquarian Society of Lancashire and Cheshire*, LXIII (1952-1953), 52-86.

Chandler, G. *Liverpool*. London, 1957.

Cotter, E.P. *The Port of Liverpool, Including Birkenhead and Garston*. Washington, DC, 1929.

Davies, P.N. "Group Enterprise: Strength and Hazards. Business History and the Teaching of Business Management." In S. Marriner (ed.). *Business and Businessmen: Studies in Business, Economic and Accounting History*. Liverpool, 1978, 141-168.

Deane, H.J. "The Development of Cranagc Facilitics for Discharging Vessels of the Largest Size." *Proceedings of the Institution of Civil Engineers Conference,* 1921, section V, 21-24.

Dickinson, R. "James Nasmyth and the Liverpool Iron Trade." *Transactions of the Historic Society of Lancashire and Cheshire*, CVIII (1956), 83-104.

Dixon, D.F. "Petrol Distribution in the United Kingdom, 1900-1915." *Business History,* VI (1963), 1-19.

Doherty, W.J. "Description of Cofferdams used at Dublin, Birkenhead and Hull." *Minutes of the Proceedings of the Institution of Civil Engineers*, LI (1877-1878), 137-160.

Dyos, H.J.and Aldcroft, D.H. *British Transport. An Economic Survey from the Seventeenth Century to the Twentieth*. Leicester, 1969.

Fischer, L.R. "Maritime Infrastructure." In J. Armstrong and A. Kunz (eds.). *Coastal Shipping in the European Economy 1750-1980*. Mainz, 2002.

Fischer, L.R. and Jarvis, A. (eds.). *Harbours and Havens: Essays in Honour of Gordon Jackson*. St. Johns, 1999.

Flayhart, W.H., *The American Line, 1871-1902*. New York, 2000.

Gibson, A.H. "Tidal and River Models." *Journal of the Institution of Civil Engineers*, VIII (1935-1936), 699-722.

Great Britain. Department of Scientific and Industrial Research. *Water Pollution Research, Technical Paper No. 7: Estuary of the River Mersey*. London, 1938.

Great Britain. Parliament. *Parliamentary Papers*. Departmental Committee on Petroleum Spirit, 1910 (Cd 5175), XLIV.609.

________. ______. ______. Royal Commission on Railways, 1867 (Cd 3844), XXXVIII part.1.

________. ______. ______. Royal Commission on the Natural Resources, Trade and Legislation of Certain Portions of His Majesty's Dominions, 1917-1918 (Cd 8641), IX.523.

________. ______. ______. Royal Commission on the Port of London, 1902 (Cd 1151), XLIII.222.

________. ______. ______. Select Committee on Transport, 1918 (130, 136) IV. 465.

Grieves, K.R. "The Liverpool Dock Battalion: Military Intervention in the Mersey Docks, 1915-1918." *Transactions of the Historic Society of Lancashire and Cheshire*, CXXXI (1981), 140-158.

Griffith, Sir John Purser. *Report on the Proposed Training Banks along the Crosby Channel: Port of Liverpool*. Liverpool, 1922.

Handbook for the Conference of the Institute of Public Health. Liverpool, 1903.

Harlaftis, G. *A History of Greek-owned Shipping. The Making of an International Tramp Fleet, 1830 to the Present*. London, 1996.

Holt, A. (ed.). *Merseyside. A Handbook to Liverpool and District*. Liverpool, 1923.

Holzapfel, A.C.A. "Ships' Composition." *Transactions of the Institution of Naval Architects*, XLVI (1904), 252-272.

Hyde, F.E. *Liverpool and the Mersey: An Economic History of a Port, 1700-1970*. Newton Abbot, 1971.

Jackson, G. *The History and Archaeology of Ports*. Tadworth, 1983.

Jarvis, A. *Liverpool Central Docks 1799-1905. An Illustrated History*. Stroud, 1991.

_______. *The Liverpool Dock Engineers*, Stroud, 1997.

_______. "G.F. Lyster and the Role of the Dock Engineer 1861-97." *Mariner's Mirror,* LXXVIII (1992), 177-199.

_______. "The Port of Liverpool and the Shipowners." *The Great Circle,* XVI (1994), 1-22.

_______. "Attempts at Reducing the Expenditure of the Port of Liverpool 1836-1913." In P.C. van Royen, *et al.* (eds.). *Frutta di Mare: Evolution and Revolution in the Maritime World in the 19th and 20th Centuries*. Amsterdam, 1998.

_______. "A.G. Lyster's *Report on the Several Methods of Dealing with Grain in the United States, and also on the Elevated Railroads of New York*, 1880." *International Journal of Maritime History*, X (1998), 155-182.

_______. "The Mersey Docks & Harbour Board in Anglesey: Dinorben Quarry." *Maritime Wales*, XIX (1997-1978) 93-99.

_______. "Protection and Decoration: A Tentative Investigation into Painting Ships before the Great War." *The Great Circle*, XXII (2000), 24-37.

_______. "Corruption in the Port of Liverpool." forthcoming.

_______. "Bulk Petroleum Importation into Liverpool before 1930." In A. Jarvis (ed.). *Foul Berths and French Spies. Essays on the Port of Liverpool, c. 1800-1930*. Liverpool, 2003, 46-55.

_______. "Port History: Some Thoughts on Where it Came from and Where it Might be Going." In L.R. Fischer and A. Jarvis (eds.). *Harbours and Havens: Essays in Honour of Gordon Jackson*. St. Johns, 1999.

_______. "The Members of the Mersey Docks & Harbour Board and Their Way of Doing Business." *International Journal of Maritime History*, VI (1994), 122-139.

_______. "An Historical Backwater: The Fishing and Fish-trading of Liverpool." *Northern Seas Yearbook (*1995), 51-75.

_______. "Managing Change: The Organisation of Port Authorities at the Turn of the Twentieth Century." *The Northern Mariner/Le Marin du nord*, VI, No. 2 (April 1996), 31-42.

_______. "Forgotten Heroes: The Dock Construction Workers." *Northern Seas Yearbook* (1997), 77-98.

_______. "The Port of Liverpool and the Shipowners, c. 1910-38." *The Northern Mariner/Le Marin du nord*, XII (2002), 23-40.

_______. "Land Policies in the Port of Liverpool, 1857-1930." *International Journal of Maritime History*, XIV, No. 1 (June 2002), 115-133.

_______. "The Fleet of the Mersey Docks & Harbour Board." *Northern Seas Yearbook*, forthcoming.

_______. "Think Big? Returns on Different Scales of Investment in the Port of Liverpool, 1873-1905." In A. Jarvis (ed.). *Foul Berths and French Spies. Essays on the Port of Liverpool, c. 1800-1930*. Liverpool, 2003, 20-34.

_______ and Smith, K. (eds.). *Albert Dock: Trade and Technology*, Liverpool, 1999.

Jones, W.H. "The Brocklebank Dock Improvements on the Mersey Dock Estate." *Transactions of the Liverpool Engineering Society,* XXXIII (1912), 164-200.

_______. "Gladstone Graving Dock, Liverpool." *Transactions of the Liverpool Engineering Society*, XXXVIII (1916-1917), 199-210.

Jeremy, D.J. and Farnie, D.A. "The Ranking of Firms, the Counting of Employees and the Classification of Data: A Cautionary Note." *Business History*, XLIII (2001), 105-118.

Kellett, J.R. *The Impact of Railways on Victorian Cities*. London, 1969.

Kirkaldy, A.W. *British Shipping: Its History, Organisation and Importance*. London, 1914; reprint, Newton Abbot, 1970.

Knightley, P. *The Rise and Fall of the House of Vestey. The True Story of How Britain's Richest Family Beat the Taxman – and Came to Grief*. London, 1993.

Large, F. *Faster than the Wind: A History of and a Guide to the Liverpool to Holyhead Telegraph*. Liverpool, 1998.

Lee, C.E. *The Blue Riband: The Romance of the Atlantic Ferry*. London, n.d. [1930].

Liverpool Steam Ship Owners' Association. *Report on Liverpool Port Dues on Shipping*. Liverpool, 1931.

_______. *Report of the Liverpool Steam Ship Owners' Association*. Liverpool, 1955.

Longbottom, K. *Liverpool and the Mersey. Vol. 1: Gladstone Dock and the Great Liners*. Peterborough, 1995.

Lyster, A G. "Works Recently Carried Out and in Contemplation at the Port of Liverpool." *VIII International Congress on Navigation*. Paris, 1900, 1-10.

_______. "The System of Sluices at the New North Works, Canada Basin, Liverpool." *Transactions of the Liverpool Engineering Society*, I (1876-1877), 13-20.

_______. "Sand-Pump Dredgers." *Minutes of the Proceedings of the Institution of Civil Engineers,* CXXXVIII (1899), 394-398.

_______. "Presidential Address." *Minutes of the Proceedings of the Institution of Civil Engineers*, CXXV (1913-1914), 4-41.

Lyster, G.F. "Recent Dock Extensions at Liverpool." *Minutes of the Proceedings of the Institution of Civil Engineers*, C (1889-1890), 2-114.

McCarron, K. *Meat at Woodside. The Birkenhead Livestock Trade 1878-1981*. Birkenhead, 1991.

McCart, N. *Atlantic Liners of the Cunard Line. From 1884 to the Present Day*. Wellingborough, 1990.

Marquis, F.J. (ed.). *Handbook of Employments in Liverpool*. Liverpool, 1916.

Matthews, Sir William. "Presidential Address." *Minutes of the Proceedings of the Institution of Civil Engineers*, CLXXI (1907-1908), 5-48.

Milne, G.J. *Trade and Traders in Mid-Victorian Liverpool: Mercantile Business and the Making of a World Port*. Liverpool, 2000.

_______. "Port Politics: Interest, Faction and Port Management in Mid-Victorian Liverpool." in L.R. Fischer and A. Jarvis (eds.). *Harbours and Havens: Essays in Honour of Gordon Jackson*. St. John's, 1999.

_______ and Tonks, G. "Specialised Port Facilities on Trial: Liverpool's Albert Dock in the Nineteenth Century." In A. Jarvis and K. Smith (eds.). *Albert Dock: Trade and Technology*, Liverpool, 1999.

Mitchell, B.R. *British Historical Statistics*. Cambridge, 1988.

Moss, W. *The Liverpool Guide, Including a Sketch of the Environs*. Liverpool, 1796; reprint, Liverpool, 1974.

Mountfield, S. *Western Gateway: A History of the Mersey Docks and Harbour Board*. Liverpool, 1965.

Mowat, M. "Some Recent Grain-handling and Storing Appliances at the Millwall Docks." *Minutes of the Proceedings of the Institution of Civil Engineers*, CLXXVII (1908-1909), 59-92.

Muir, J.R.B. *A History of Liverpool*. London, 1907.

Musson, A.E. *Enterprise in Soap and Chemicals. Joseph Crosfield and Sons, Limited, 1815-1965*. Manchester, 1965.

Neal, F. *Sectarian Violence: The Liverpool Experience. An Aspect of Anglo-Irish History*. Manchester, 1989.

Nevett, T.R. *Advertising in Britain. A History*. London, 1982.

O'Mara, P. *The Autobiography of a Liverpool Irish Slummy*. Liverpool, 1934; reprint, Liverpool, 1972.

Oglethorpe, W.A. "Gladstone Graving Dock, Liverpool - Machinery." *Transactions of the Liverpool Engineering Society*, XXXVIII (1916-1917), 211-246.

_______. "Modern Dredging Appliances." *Transactions of the Liverpool Engineering Society*, XXXVI (1909-1910), 27-56.

Port Facilities of Great Britain: The Report of the Port Facilities Committee appointed by the Chamber of Shipping, by the Association of British Chambers of Commerce and by the Federation of British Industries. London, 1924.

Ramsbotham, J.F. "Graving Dock Construction, Dealing with the New Brocklebank Graving Dock." *Transactions of the Liverpool Engineering Society*, XXVIII (1907), 45-70 and 101-116.

_______. "Demolition of Old Masonry and Concrete." *Transactions of the Liverpool Engineering Society*, XXIX (1908-1909), 22-56.

Report of Committee Appointed by The Mersey Docks and Harbour Board to Investigate the Effects of the Discharge of Crude Sewage into the River Mersey. Liverpool, 1930.

Report of the Meeting of the Deputation from the Joint Committee of the City Council, Dock Board and Mercantile Associations with the Directors of the Railway Companies Who Have Termini in Liverpool. Liverpool, 1881.

Saul, S.B. *The Myth of the Great Depression, 1873-1896*. London, 1969.

Shapton, W. "Description of the Warehouse and Machinery for the Storage and Transit of Grain at the Alexandra Dock, Liverpool." *Proceedings of the Institution of Mechanical Engineers* (1891), 372-385.

Simey, M. *The Disinherited Society. A Personal View of Social Responsibility in Liverpool during the Twentieth Century*. Liverpool, 1996.

Simmons, J. *The Victorian Railway*. London, 1995.

Smith, A. *The Illustrated Guide to Liverpool Herculaneum Pottery 1796-1840*. London, 1970.

Skempton, A.W. "A History of the Steam Dredger, 1797-1830." *Transactions of the Newcomen Society*, XLVII (1974-1976), 97-116.

Stammers, M.K. *Mersey Flats and Flatmen*. Liverpool, 1993.

_______. "A Dock Where Nothing Will Ever Be Done: Albert Dock in the Twentieth Century." In A. Jarvis and K. Smith (eds.). *Albert Dock: Trade and Technology*. Liverpool, 1999.

Starkey, D.J., (ed.). *Shipping Movements in the Ports of the United Kingdom 1871-1913: A Statistical Profile*. Exeter 1999.

_______ and Jamieson, A.G. (eds.). *Exploiting the Sea: Aspects of Britain's Maritime Economy since 1870*. Exeter, 1998.

Stevens, R.W. *On the Stowage of Ships and Their Cargoes*. London, 1869.

Sunday at Home. 1896, 163-168, 246-251, 307-312, 441-447.

Taplin, E.L. *Near to Revolution: The Liverpool General Transport Strike of 1911*. Liverpool, 1994.

Tarn, J.N. *Working Class Housing in 19th-Century Britain*. London, 1971.

_______. *Five Per Cent Philanthropy. An Account of Housing in Urban Areas between 1840 and 1914*. Cambridge, 1973.

Taylor, F.M. du Plat. *The Design, Construction and Maintenance of Docks, Wharves and Piers*. London, 1928.

Vernon-Harcourt, L.F. "The Milan Navigation Congress of 1905." Unpublished manuscript, Institution of Civil Engineers, 1905.

Waller, P.J. *Democracy and Sectarianism: A Political and Social History of Liverpool 1868-1939*. Liverpool, 1981.

Walton, J.K. and Wilcox, A. (eds.). *Low Life and Moral Improvement in Mid-Victorian Liverpool. Liverpool through the Journalism of Hugh Shimmin*. Leicester, 1991.

Wardley, P. "The Emergence of Big Business: The Largest Corporate Employers of Labour in the United Kingdom, Germany and the United States c. 1907." *Business History*, XLI (1999), 88-116.

_______. "On the Ranking of Firms: A Response to Jeremy & Farnie." *Business History*, XLIII (2001), 119-134.

Warner, L.A.P. "The Mersey Dock Estate – Its Management and Principal Trades." *Journal of the Institute of Transport,* Special Liverpool Congress Number (1928), 443-460.

White, Sir William. "Presidential Address." *Minutes of the Proceedings of the Institution of Civil Engineers*, CLV (1903-1904), part 1, 2-169.

Williams, D.M. "State Regulation of Merchant Shipping 1839-1914: The Bulk Carrying Trades." In S. Palmer and G. Williams (eds.). *Charted and Uncharted Waters*. London, 1982.

Index